CONTEMPORARY

# NUMBER POWER

A REAL WORLD APPROACH TO MATH

## Fractions, Decimals, and Percents

JERRY HOWETT

*Bothell, WA • Chicago, IL • Columbus, OH • New York, NY*

www.mheonline.com

Cover 7 133 Ryan McKay/Photodisc/Getty Images.



Send all inquiries to:
Contemporary/McGraw-Hill
130 E. Randolph, Suite 400
Chicago, IL 60601

ISBN: 978-0-07-659227-2
MHID: 0-07-659227-8

Printed in the United States of America.

8 9 10 RHR 20 19 18 17 16

# TABLE OF CONTENTS

## Percents

# TO THE STUDENT

Welcome to *Fractions, Decimals, and Percents:*

This book is designed to help you build and use the fraction, decimal, and percent skills found in the workplace and everyday experiences. This book extends the whole number skills developed in *Number Power: Addition, Subtraction, Multiplication, and Division.*

The first section of the book, Building Number Power, provides easy-to-follow instructions and plenty of practice working with fractions, decimals, and percents as well as practice using estimation in problem solving. Each chapter begins with a skills inventory so that you can evaluate your strengths and weaknesses. This section ends with a review so that you can measure your progress.

The second section of the book, Using Number Power, gives you an opportunity to use your math skills to solve real-life problems.

Learning how and when to use a calculator is an important skill to develop. In real life, problem solving often involves the smart use of a calculator – especially when working with large numbers. You need to know when an answer on a calculator makes sense, so be sure to have an estimated answer in mind. Remember that a calculator can help you only if you set up the problem and use the calculator correctly.

At the back of the book is an answer key for all the exercises and tests. Checking the answer key after you work through a lesson will help you measure your progress. Also in the back are a list of the measurements and formulas used in the book and a glossary of mathematical terms. Inside the back cover is a chart to help you keep track of your score on each exercise.

After carefully working through this book, you will be able to handle the fractions, decimals, and percents needed in daily living, in school, in the workplace, and on many standardized tests, including the GED. The skills you develop will give you a solid foundation for your increasing number power.

# Pretest

This test will tell you which sections of the book you need to concentrate on. Do every problem that you can. Correct answers are listed by page number at the back of the book. After you check your answers, the chart at the end of the test will guide you to the pages of the book where you need work.

**1.** Reduce $\frac{18}{24}$ to the lowest terms.

**2.** $2\frac{3}{4} + 4\frac{5}{6} + 1\frac{3}{8} =$

**3.** Is the sum of $\frac{4}{7} + \frac{8}{15}$ greater than 1, less than 1, or equal to 1?

**4.** Sam does home renovations. He is building new kitchen cabinets for a customer. The base of the cabinet is $34\frac{1}{4}$ inches high. The countertop, which rests on the base, is $1\frac{7}{8}$ inches thick. Find the height of the cabinet including the countertop.

**5.** $12\frac{3}{10} - 7\frac{2}{3} =$

**6.** Which of the following gives the best whole number estimate of $9\frac{1}{4} - 6\frac{7}{8}$?

**a.** $10 - 7 =$ **b.** $9 - 7 =$ **c.** $9 - 6 =$

**7.** From a 5-pound bag of flour, Maura used $2\frac{3}{16}$ pounds to bake cakes. Find the weight of the flour left in the bag.

**8.** $\frac{5}{8} \times \frac{4}{9} =$

**9.** $2\frac{2}{3} \times 1\frac{3}{4} \times 7\frac{1}{2} =$

**10.** In the problem $2\frac{7}{8} \times 5\frac{1}{6}$, first round each number to the nearest whole number. Then multiply.

11. Because of snow, Mavis's evening Spanish class was canceled. The next week, $\frac{2}{3}$ of the 24 students in the class agreed to have a make-up class. How many students wanted the make-up class?

12. $4\frac{1}{2} \div 9 =$

13. $3\frac{1}{4} \div 1\frac{1}{8} =$

14. Jake cut a pipe that was $17\frac{1}{2}$ inches long into five equal pieces. Assuming there was no waste, how long was each piece?

15. On a Spanish vocabulary quiz, Connie got $\frac{3}{4}$ of the words right. She answered 36 words correctly. How many words were on the quiz?

16. Write twelve and sixteen thousandths as a mixed decimal.

17. Order the following decimals from smallest to largest: 0.057, 0.7, 0.075, 0.57

18. $5.63 + 12 + 1.078 =$

19. A carpenter attached a metal cover that was 0.125 inch thick onto a board that was 1.5 inches thick. Find the combined thickness of the board and the cover.

20. $2.8 - 1.097 =$

**21.** In the problem $8.37 - 1.629$, first round each number to the nearest tenth. Then subtract.

**22.** The area of Washington, D.C. is 61.4 square miles. The area of St. Louis is 66.2 square miles. The area of St. Louis is how much larger than the area of Washington, D.C.?

**23.** $4.07 \times 2.6 =$

**24.** In the problem $89.6 \times 0.78$, first round 89.6 to the nearest *ten* and round 0.78 to the nearest *tenth*. Then multiply.

**25.** To the nearest penny, what is the price of 3.2 pounds of ham that costs $3.99 per pound?

**26.** $11.02 \div 2.9 =$

**27.** $19.3 \div 100 =$

**28.** $90 \div 0.45 =$

**29.** Find the answer to $8.8 \div 0.7$ to the nearest tenth.

**30.** Three crates weigh a total of 128.6 pounds. To the nearest tenth, what is the average weight of each crate?

**31.** Write 4.8% as a decimal.

**32.** Write $\frac{5}{12}$ as a percent.

**33.** Write $6\frac{1}{4}\%$ as a fraction in lowest terms.

**34.** Find 27% of 300.

**35.** Find $12\frac{1}{2}\%$ of 48.

**36.** Which of the following is the same as 20% of 75?

**a.** $75 \div 2$
**b.** $75 \div 3$
**c.** $75 \div 4$
**d.** $75 \div 5$

**37.** Iris works in the quality control department of a factory that makes children's clothes. In a week when the factory produced 350 items, Iris found 4% of them to be defective. How many clothing items were defective that week?

**38.** Marlene is in charge of ticket sales at a county fair. The attendance on opening night was 2,400 people. The attendance on the last night of the fair was 15% more than on the opening night. Find the attendance for the last night of the fair.

**39.** 84 is what percent of 112?

**40.** John wants to buy an outdoor grill that costs $400. So far he has saved $250. What percent of the cost of the grill has he saved?

**41.** When a home delivery meal service started, they prepared and delivered 320 meals. One year later the service was delivering 576 meals. By what percent did the number of meals increase?

**42.** In December, Yanni's Chocolate Shop received 180 telephone orders. In January, they received only 72 telephone orders. By what percent did the number of telephone orders decrease from December to January?

**43.** 60% of what number is 72?

**44.** On a flight to Chicago, there were 104 occupied seats. The occupied seats represent 80% of the total number of seats on the plane. How many seats are on the plane?

**45.** The road crew in the village of Pleasant Hill hopes to have every street in the village resurfaced by the end of October. By the first of October, the crew had resurfaced 36 streets, which represent 75% of the total. How many streets are there in Pleasant Hill?

# PRETEST CHART

If you miss more than one problem in any section of this test, you should complete the lessons on the practice pages indicated on this chart. If you miss only one problem in a section of a test, you may not need further study in that section. However, before you skip those lessons, we recommend that you complete the review test at the end of that chapter. For example, if you miss one problem about fractions, you should pass the Fractions Review (page 60) before beginning the chapter on decimals. This longer inventory will be a more precise indicator of your skill level.

| Problem Numbers | Skill Area | Practice Pages |
|---|---|---|
| 1, 2, 3, 4 | understanding and adding fractions | 11–30 |
| 5, 6, 7 | subtracting fractions | 31–39 |
| 8, 9, 10, 11 | multiplying fractions | 40–47 |
| 12, 13, 14, 15 | dividing fractions | 48–59 |
| 16, 17, 18, 19 | understanding and adding decimals | 64–76 |
| 20, 21, 22 | subtracting decimals | 77–80 |
| 23, 24, 25 | multiplying decimals | 81–87 |
| 26, 27, 28, 29, 30 | dividing decimals | 88–96 |
| 31, 32, 33 | understanding percents | 101–106 |
| 34, 35, 36, 37, 38 | finding a percent | 107–112 |
| 39, 40, 41, 42 | finding what percent one number is of another | 113–118 |
| 43, 44, 45 | finding a number when a percent is given | 119–122 |

# BUILDING NUMBER POWER

# FRACTIONS

## Fraction Skills Inventory

This inventory will tell you whether you need to work through the fractions section of this book. Do all the problems that you can. There is no time limit. Work carefully and check your answers, but do not use outside help. Correct answers are listed by page number at the back of the book.

**1.** A pound contains 16 ounces. 10 ounces is what fraction of a pound?

**2.** Which fractions in this list are greater than $\frac{1}{2}$? $\frac{5}{6}, \frac{4}{8}, \frac{3}{4}, \frac{9}{20}$

**3.** Reduce $\frac{28}{42}$ to lowest terms.

**4.** Write $\frac{24}{10}$ as a mixed number.

**5.** Write $5\frac{3}{7}$ as an improper fraction.

**6.** Change $\frac{2}{3}$ to eighteenths.

**7.** $3\frac{11}{15} + 4\frac{7}{15} =$

**8.** $3\frac{1}{3} + 2\frac{3}{4} + 5\frac{5}{6} =$

**9.** Without finding the exact sum, tell whether $\frac{5}{12} + \frac{1}{2}$ is less than 1, greater than 1, or equal to 1.

**10.** A wooden crate weighing $2\frac{5}{16}$ pounds contains grapefruit weighing $24\frac{1}{2}$ pounds. What is the combined weight of the crate and the grapefruit?

**11.** Fred is a traveling salesman. He spent $1\frac{3}{4}$ hours driving on Monday, $2\frac{1}{3}$ hours driving on Tuesday, and $2\frac{1}{6}$ hours driving on Wednesday. Find the total number of hours he drove those three days.

**12.** $12\frac{3}{4} - 6\frac{2}{5} =$

**13.** $9\frac{1}{8} - 4\frac{5}{8} =$

**14.** $10\frac{2}{7} - 6\frac{2}{3} =$

**15.** Maxine weighed 135 pounds. By dieting she lost $10\frac{1}{2}$ pounds. How much did she weigh after her diet?

**16.** From a piece of cloth $42\frac{1}{4}$ inches long, Celeste cut a strip $14\frac{3}{4}$ inches long. How long was the remaining piece?

**17.** $\frac{7}{10} \times \frac{3}{4} =$

**18.** $\frac{3}{5} \times \frac{10}{21} \times \frac{7}{8} =$

**19.** $2\frac{2}{5} \times 6\frac{1}{4} =$

**20.** In the problem $1\frac{3}{5} \times 3\frac{1}{4}$, round each number to the nearest whole number. Then multiply.

**21.** Bill shipped $\frac{7}{10}$ of the crates in his warehouse by air freight. If there were 40 crates in the warehouse, how many did he ship?

**22.** Gloria works $7\frac{1}{2}$ hours a day. How many hours does she work in a 5-day work week?

**23.** $\frac{5}{8} \div \frac{9}{16} =$

**24.** $12 \div \frac{3}{4} =$

**25.** $\frac{7}{9} \div 14 =$

**26.** $3\frac{3}{4} \div 1\frac{5}{7} =$

**27.** $\frac{3}{4}$ of what number is 36?

**28.** How many $\frac{1}{2}$-pound bags can be filled with 14 pounds of peanuts?

**29.** From a strip of wood 125 inches long, George is cutting pieces $12\frac{1}{2}$ inches long. How many pieces can he cut from the long strip?

## FRACTION SKILLS INVENTORY CHART

If you missed more than one problem in any group below, work through the practice pages for that group. Then redo the problems you got wrong on the Fraction Skills Inventory. If you had a passing score on all five groups of problems, redo any problem you missed and begin the Decimal Skills Inventory on page 62.

| Problem Numbers | Skill Area | Practice Pages |
|---|---|---|
| 1, 2, 3, 4, 5, 6 | understanding fractions | 11–21 |
| 7, 8, 9, 10, 11 | adding fractions | 22–30 |
| 12, 13, 14, 15, 16 | subtracting fractions | 31–39 |
| 17, 18, 19, 20, 21, 22 | multiplying fractions | 40–47 |
| 23, 24, 25, 26, 27, 28, 29 | dividing fractions | 48–59 |

homework 11, – 14

# Understanding Fractions

A **fraction** is a part of a whole. A penny is a fraction of a dollar. It is one of the 100 equal parts of a dollar or $\frac{1}{100}$ (one-hundred*th*) of a dollar. An inch is a fraction of a foot. It is one of the 12 equal parts of a foot or $\frac{1}{12}$ (one-twelf*th*) of a foot. 5 days are a fraction of a week. They are 5 of the 7 equal parts of a week or $\frac{5}{7}$ (five-seven*ths*) of a week.

The two numbers in a fraction are called the **numerator** and the **denominator.**

| | |
|---|---|
| **numerator** | tells how many parts you have |
| **denominator** | tells how many parts in the whole |

EXAMPLE The fraction $\frac{3}{4}$ tells you what part of the figure at the right is shaded. 3 parts are shaded. The whole figure is divided into 4 equal parts.

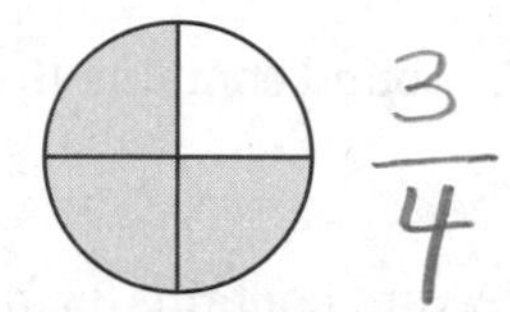

$\frac{3}{4}$

**Write fractions that represent the part of each figure that is shaded.**

**1.** 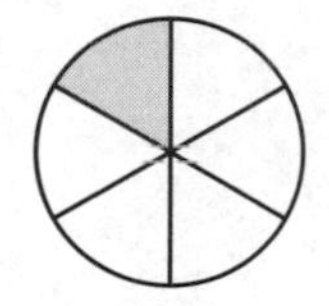 $\frac{1}{6}$

**2.** 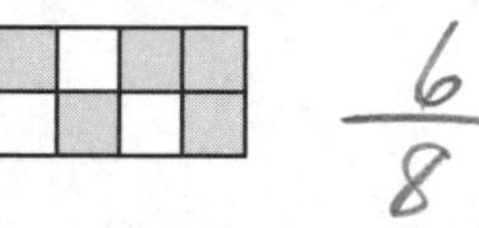 $\frac{6}{8}$

**3.** 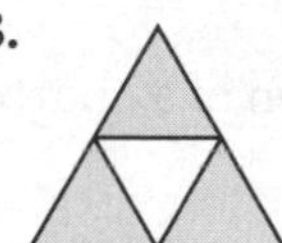 $\frac{3}{4}$

**4.** 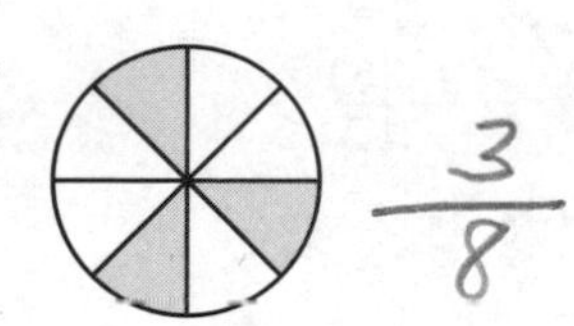 $\frac{3}{8}$

**5.** 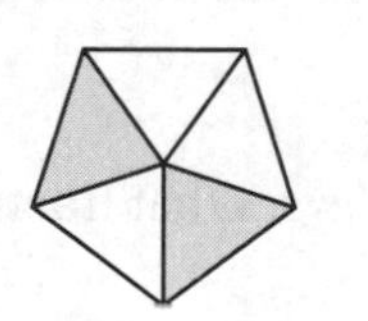 $\frac{2}{5}$

**6.** 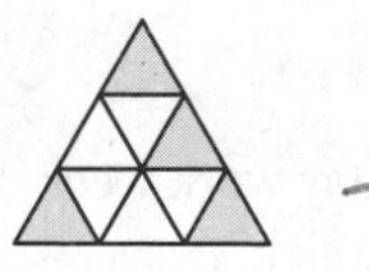 $\frac{4}{9}$

**7.** 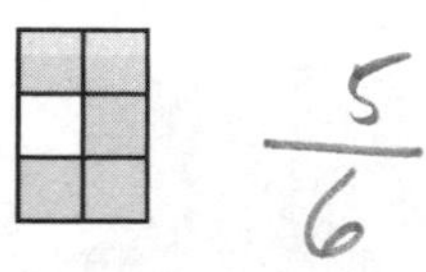 $\frac{5}{6}$

**8.** 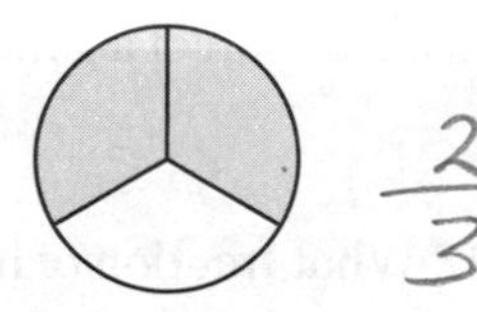 $\frac{2}{3}$

**9.** 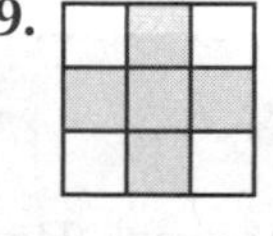 $\frac{5}{9}$

# Writing Fractions

Since there are three feet in one yard, one foot is $\frac{1}{3}$ of a yard. Two feet make up $\frac{2}{3}$ of a yard.

**Write fractions for each of the *parts* described below.**

1. A foot contains 12 inches. 5 inches is what fraction of a foot?

2. 47¢ is what fraction of a dollar?

3. A pound contains 16 ounces. 9 ounces is what fraction of a pound?

4. A yard contains 36 inches. 23 inches is what fraction of a yard?

5. 7 months is what fraction of a year?

6. 8¢ is what fraction of a quarter?

7. Earlene wants a jacket that costs $60. She has saved $43. What fraction of the amount that she needs has she saved?

8. There are 2,000 pounds in a ton. 1,351 pounds is what fraction of a ton?

9. During a 5-day work week, Pete was sick for 2 days. What fraction of the work week was he sick?

10. There are 100 centimeters in a meter. 63 centimeters is what fraction of a meter?

11. There are 4 quarts in a gallon. 3 quarts is what fraction of a gallon?

12. David makes $720 a week. He has spent $217. What fraction of his week's pay has he spent?

13. Madge has typed 77 pages of a report that contains 280 pages. What fraction of the report has she typed?

# Identifying Forms of Fractions

In a **proper fraction,** the top number is *less than* the bottom number.

EXAMPLES $\frac{1}{3}, \frac{3}{10}, \frac{7}{19}$

A proper fraction is less than all the parts the whole is divided into. The value of a proper fraction is *always less than one.*

In an **improper fraction,** the top number is *equal to or larger than* the bottom number.

EXAMPLES $\frac{3}{2}, \frac{9}{4}, \frac{8}{8}$

An improper fraction is all the parts that a whole is divided into, such as $\frac{8}{8}$, or it is more than the total parts in the whole. The value of an improper fraction is either equal to one or more than one.

In a **mixed number,** a whole number is written next to a proper fraction.

EXAMPLES $1\frac{2}{5}, 3\frac{1}{2}, 10\frac{4}{7}$

---

**Tell whether each of the following is a proper fraction (P), an improper fraction (I), or a mixed number (M).**

**1.** $\frac{9}{6}$ $\quad\frac{8}{30}$ $\quad 4\frac{1}{2}$ $\quad\frac{10}{10}$

**2.** $\frac{15}{16}$ $\quad\frac{20}{19}$ $\quad\frac{17}{17}$ $\quad 12\frac{1}{10}$

**3.** $9\frac{2}{7}$ $\quad\frac{55}{100}$ $\quad\frac{2}{200}$ $\quad\frac{200}{2}$

**4.** $\frac{75}{110}$ $\quad\frac{110}{75}$ $\quad 1\frac{8}{9}$ $\quad\frac{8}{50}$

# Thinking About the Size of Fractions

In a proper fraction, the **numerator** (top number) is less than the **denominator** (bottom number). The size of the numerator *compared* to the size of the denominator tells you something about the size of the fraction.

> A fraction is equal to one-half when the numerator is exactly half of the denominator.

**EXAMPLES** The fractions $\frac{2}{4}$, $\frac{5}{10}$, and $\frac{15}{30}$ are each equal to one-half.

**1.** Circle the fractions that are *equal to* $\frac{1}{2}$ in the following list.

$\frac{4}{9}$ $\quad\frac{6}{12}$ $\quad\frac{9}{18}$ $\quad\frac{7}{15}$ $\quad\frac{3}{7}$ $\quad\frac{8}{16}$ $\quad\frac{12}{24}$

> A fraction is less than one-half when the numerator is less than half of the denominator.

**EXAMPLES** The fractions $\frac{1}{3}$, $\frac{2}{5}$, and $\frac{12}{50}$ are each less than one-half.

**2.** Circle the fractions that are *less than* $\frac{1}{2}$ in the following list.

$\frac{2}{7}$ $\quad\frac{3}{8}$ $\quad\frac{6}{12}$ $\quad\frac{9}{10}$ $\quad\frac{4}{9}$ $\quad\frac{9}{20}$ $\quad\frac{4}{15}$

> A fraction is greater than one-half when the numerator is more than half of the denominator.

**EXAMPLES** The fractions $\frac{3}{5}$, $\frac{7}{10}$, and $\frac{24}{35}$ are each greater than one-half.

**3.** Circle the fractions that are *greater than* $\frac{1}{2}$ in the following list.

$\frac{3}{4}$ $\quad\frac{3}{6}$ $\quad\frac{7}{8}$ $\quad\frac{9}{15}$ $\quad\frac{8}{12}$ $\quad\frac{10}{20}$ $\quad\frac{17}{30}$

**The symbol $=$ means "is equal to."** $\quad\frac{12}{24} = \frac{1}{2}$ because 12 is half of 24.

**The symbol $<$ means "is less than."** $\quad\frac{5}{14} < \frac{1}{2}$ because 5 is less than half of 14.

**The symbol $>$ means "is greater than."** $\quad\frac{9}{10} > \frac{1}{2}$ because 9 is more than half of 10.

**In the box between each pair of fractions, write the symbol that makes the statement true. The first problem is done as an example.**

**4.** $\frac{3}{5} \boxed{>} \frac{1}{2}$ $\qquad\frac{8}{16} \square \frac{1}{2}$ $\qquad\frac{7}{20} \square \frac{1}{2}$ $\qquad\frac{9}{15} \square \frac{1}{2}$

homework

# Reducing Fractions

A quarter (25¢) represents 25 pennies out of the total of 100 pennies in a dollar. A quarter is $\frac{25}{100}$ of a dollar. A quarter also equals five nickels out of the total of 20 nickels in a dollar. A quarter is $\frac{5}{20}$ of a dollar. A quarter is also one of the four equal parts of a dollar or $\frac{1}{4}$ of a dollar. The fraction $\frac{1}{4}$ is the *reduced* form of both $\frac{25}{100}$ and $\frac{5}{20}$.

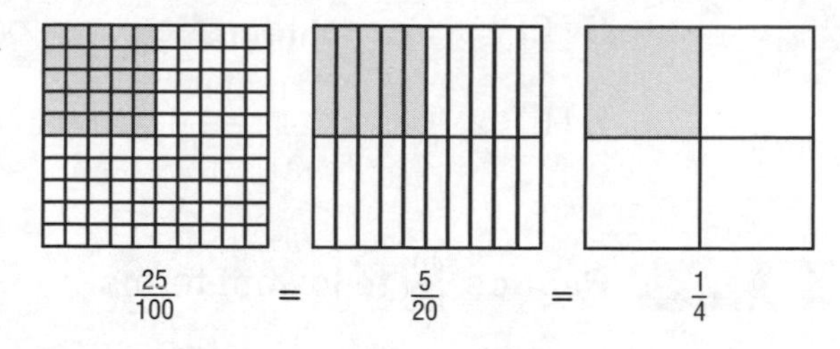

$\frac{25}{100} = \frac{5}{20} = \frac{1}{4}$

**Reducing a fraction** means writing the fraction with smaller numbers without changing the value of the fraction. Study the following examples to see how fractions are reduced.

EXAMPLE 1 Reduce $\frac{15}{20}$.

STEP 1 Find a number that divides evenly into both the numerator and the denominator. 5 divides evenly into both 15 and 20.

$$\frac{15 \div 5}{20 \div 5} = \mathbf{\frac{3}{4}}$$

STEP 2 Check to see if another number divides evenly into both 3 and 4. Since no other number besides 1 divides evenly into 3 and 4, the fraction is reduced as far as it will go.

EXAMPLE 2 Reduce $\frac{32}{64}$.

STEP 1 Find a number that divides evenly into both the numerator and the denominator. 8 divides evenly into both 32 and 64.

$$\frac{32 \div 8}{64 \div 8} = \mathbf{\frac{4}{8}}$$

STEP 2 Check to see if another number divides evenly into both 4 and 8. 4 divides evenly into both 4 and 8.

$$\frac{4 \div 4}{8 \div 4} = \mathbf{\frac{1}{2}}$$

STEP 3 Check to see if another number divides evenly into both 1 and 2. Since no other number besides 1 divides evenly into 1 and 2, the fraction is reduced as far as it will go.

When you reduce a fraction, the *value* of the fraction does not change. A reduced fraction is *equal to* the original fraction. For example, $\frac{32}{64} = \frac{1}{2}$. When a fraction is reduced as much as possible, the fraction is reduced to **lowest terms.**

When both the numerator and the denominator of a fraction end with zeros, you can cancel the zeros one-for-one. This is a shortcut for reducing by ten. Always check to see if you can continue to reduce.

**EXAMPLE 3** **Reduce $\frac{20}{30}$ to lowest terms.**

**STEP 1** Cancel the zeros one-for-one. $\frac{2\cancel{0}}{3\cancel{0}}$

**STEP 2** Check to see if you can continue to reduce. The fraction $\frac{2}{3}$ is reduced to lowest terms. $\mathbf{\frac{2}{3}}$

**EXAMPLE 4** **Reduce $\frac{40}{100}$ to lowest terms.**

**STEP 1** Cancel the zeros one-for-one. Be sure to cancel only one zero in the denominator since you canceled only one zero in the numerator. $\frac{4\cancel{0}}{10\cancel{0}}$

**STEP 2** Check to see if you can continue to reduce. Divide both 4 and 10 by 2. $\frac{4 \div 2}{10 \div 2} = \mathbf{\frac{2}{5}}$

---

**Reduce each fraction to lowest terms.**

**1.** $\frac{6}{12} = \frac{1}{2}$ $\quad \frac{7}{28} = \frac{1}{4}$ $\quad \frac{3}{9} = \frac{1}{3}$ $\quad \frac{9}{45} = \frac{1}{5}$ $\quad \frac{6}{48} = \frac{1}{8}$

**2.** $\frac{25}{30} = \frac{5}{6}$ $\quad \frac{32}{36} = \frac{16}{18} = \frac{8}{9}$ $\quad \frac{16}{18} = \frac{8}{9}$ $\quad \frac{21}{24} = \frac{7}{8}$ $\quad \frac{14}{21} =$

**3.** $\frac{20}{50} = \frac{10}{25} = \frac{2}{5}$ $\quad \frac{30}{90} = \frac{10}{30} = \frac{5}{15} = \frac{1}{3}$ $\quad \frac{70}{200} = \frac{35}{100}$ $\quad \frac{90}{140} = \frac{45}{70}\ \frac{9}{15}$ $\quad \frac{80}{170} = \frac{40}{85}$

**4.** $\frac{33}{77} = \frac{3}{7}$ $\quad \frac{45}{60} = \frac{9}{12} = \frac{3}{4}$ $\quad \frac{18}{36} = \frac{9}{18} = \frac{3}{6} = \frac{1}{2}$ $\quad \frac{42}{56} = \frac{21}{28}$ $\quad \frac{48}{64} = \frac{24}{32} = \frac{12}{16} = \frac{6}{8} = \frac{3}{4}$

**5.** $\frac{75}{80} =$ $\quad \frac{42\cancel{0}}{48\cancel{0}} = \frac{21}{24}$ $\quad \frac{72}{90} = \frac{36}{45}$ $\quad \frac{26}{39} =$ $\quad \frac{18}{32} = \frac{9}{16}$

**6.** $\frac{25}{50} = \frac{5}{10}$ $\quad \frac{14}{42} = \frac{7}{21}\ \frac{1}{3}$ $\quad \frac{4}{200} = \frac{1}{50}$ $\quad \frac{63}{81} = \frac{7}{9}$ $\quad \frac{35}{49} = \frac{5}{7}$

**For each problem, write a fraction. Then reduce.**

7. Eight inches is what fraction of a foot? (1 foot = 12 inches) $8/12 = \frac{2}{3}$

8. Sally has saved $4,500 for a used car that costs $7,500. What fraction of the total price has she saved? 3,000 $\frac{45}{75} = \frac{3}{5}$

9. Sam gets 15 vacation days each year. By June 1, Sam had used six of his vacation days. What fraction of the year's total did he use so far? $6/15 = \frac{2}{5}$

10. In the spring, Max weighed 200 pounds. Over the summer he lost 20 pounds. What fraction of his weight did he lose? $\frac{20}{200} = \frac{1}{10}$

11. Ten ounces is what fraction of a pound? (1 pound = 16 ounces) $\frac{10}{16} = \frac{5}{8}$

12. There are 21 students in Monica's math class. Fourteen of the students are women. Women make up what fraction of the class? $\frac{14}{21}$ $\frac{2}{3}$

13. For the class in problem 12, what fraction of the students are men? $\frac{12}{21} = \frac{4}{7}$

14. On a quiz with 12 problems, Evelyn got 3 problems wrong. What fraction of the problems did she get wrong? $\frac{3}{12} = \frac{1}{4}$

15. For the quiz in problem 14, what fraction of the problems did Evelyn get right? $\frac{9}{12} = \frac{3}{4}$

16. The distance from Sam's house to his job is 16 miles. On his way to work each morning, Sam drops his daughter off at school. The school is 2 miles from their house. What fraction of the total drive to his job has Sam completed when he drops off his daughter? $\frac{2}{16} = \frac{1}{8}$

17. Sheila has to type a 36-page report. When she took her coffee break, she had already typed 20 pages of the report. What fraction of the whole report has she typed? $\frac{20}{36} = \frac{10}{18} = \frac{5}{9}$

18. For the last problem, what fraction of the report does Sheila have left to type when she comes back from her coffee break? $\frac{16}{36} - \frac{8}{18} = \frac{4}{9}$

19. The Westside Community Center needs $24,000 for their summer youth program. So far they have raised $15,000. What fraction of the total cost have they raised? $\frac{15}{24} = \frac{5}{8}$

20. For the last problem, what fraction of the cost of the summer youth program does the community center need to raise? $\frac{9}{24} = \frac{3}{8}$

# Raising Fractions to Higher Terms

An important skill in the addition and subtraction of fractions is raising a fraction to **higher terms.** This is the opposite of reducing a fraction. To reduce a fraction, you must *divide* both the numerator and the denominator by the same number. To raise a fraction to higher terms, *multiply* both the numerator and the denominator by the same number.

**EXAMPLE 1** **Raise $\frac{2}{5}$ to a new fraction with a denominator of 20.**

**STEP 1** Divide the old denominator into the new one. The new denominator is 4 times the old one. $20 \div 5 = 4$

**STEP 2** Since the new denominator is 4 times the old one,multiply the old numerator by 4. $\frac{2 \times 4}{5 \times 4} = \mathbf{\frac{8}{20}}$

To check the answer, divide 8 and 20 by 4. $\frac{8 \div 4}{20 \div 4} = \frac{2}{5}$

**EXAMPLE 2** $\mathbf{\frac{4}{9} = \frac{?}{27}}$

**STEP 1** Divide the old denominator into the new one. The new denominator is 3 times the old one. $27 \div 9 = 3$

**STEP 2** Since the new denominator is 3 times the old one, multiply the old numerator by 3. $\frac{4 \times 3}{9 \times 3} = \mathbf{\frac{12}{27}}$

To check the answer, divide 12 and 27 by 3. $\frac{12 \div 3}{27 \div 3} = \frac{4}{9}$

**Raise each fraction to higher terms by filling in the missing numerator.**

**1.** $\frac{4}{5} = \frac{\quad}{30}$ $\frac{9}{10} = \frac{\quad}{20}$ $\frac{1}{6} = \frac{\quad}{18}$ $\frac{5}{8} = \frac{\quad}{32}$

**2.** $\frac{4}{7} = \frac{\quad}{35}$ $\frac{1}{2} = \frac{\quad}{36}$ $\frac{2}{3} = \frac{\quad}{21}$ $\frac{9}{11} = \frac{\quad}{66}$

**3.** $\frac{5}{9} = \frac{\quad}{45}$ $\frac{3}{4} = \frac{\quad}{44}$ $\frac{7}{12} = \frac{\quad}{60}$ $\frac{1}{3} = \frac{\quad}{45}$

**4.** $\frac{1}{2} = \frac{\quad}{50}$ $\frac{3}{8} = \frac{\quad}{40}$ $\frac{2}{9} = \frac{\quad}{72}$ $\frac{5}{6} = \frac{\quad}{42}$

# Changing Improper Fractions to Whole or Mixed Numbers

Earlier you learned that a fraction is a part of a whole. The fraction $\frac{3}{4}$ means a whole is divided into four parts and you have three of those parts.

You can also think of a fraction as a division problem. The fraction bar (—) or the slash (/) mean to divide. The fraction $\frac{3}{4}$ can mean *three divided by four.* You will learn more about dividing proper fractions when you study decimals.

The improper fraction $\frac{8}{8}$ means *eight divided by eight.*

**EXAMPLE 1** **Change $\frac{8}{8}$ to a whole number.**

Divide 8 by 8. $\frac{8}{8} = 8\overline{)8}$ → **1**

**EXAMPLE 2** **Change $\frac{20}{4}$ to a whole number.**

Divide 20 by 4. $\frac{20}{4} = 4\overline{)20}$ → **5**

When there is a remainder, write a fraction with the remainder over the denominator. The result is a mixed number.

**EXAMPLE 3** **Change $\frac{21}{9}$ to a mixed number.**

**STEP 1** Divide 21 by 9.

$$9\overline{)21} = 2\tfrac{3}{9}, \quad 21 - 18 = 3$$

**STEP 2** Write the remainder, 3, over the denominator, 9.

**STEP 3** Reduce $\frac{3}{9}$ by dividing 3 and 9 by 3. $\frac{3 \div 3}{9 \div 3} = \frac{1}{3}$

**ANSWER: $2\frac{1}{3}$**

For the last example, you can reduce the improper fraction before you change it to a mixed number. The result will be the same.

$$\frac{21 \div 3}{9 \div 3} = \frac{7}{3} \qquad 3\overline{)7} = \mathbf{2\tfrac{1}{3}}$$

Division word problems can be written as improper fractions. Remember that the amount being divided goes on top, and the divisor goes on the bottom.

**EXAMPLE 4** **A kitchen counter is 90 inches long. Find the length of the counter in feet. (1 foot = 12 inches)**

**STEP 1** Write the problem as an improper fraction. Divide 90 inches by the number of inches in one foot. $\frac{90}{12}$

**STEP 2** Divide 90 by 12. $12\overline{)90}$ → $7\frac{6}{12}$

**STEP 3** Reduce $\frac{6}{12}$ by dividing 6 and 12 by 6. $\frac{6 \div 6}{12 \div 6} = \frac{1}{2}$

**ANSWER:** $7\frac{1}{2}$

**Change each fraction to a whole number or a mixed number. Reduce any remaining fractions.**

1. $\frac{14}{8} =$ $\frac{33}{6} =$ $\frac{14}{5} =$ $\frac{30}{7} =$ $\frac{12}{3} =$

2. $\frac{30}{9} =$ $\frac{26}{8} =$ $\frac{18}{6} =$ $\frac{36}{10} =$ $\frac{16}{8} =$

3. $\frac{13}{12} =$ $\frac{45}{9} =$ $\frac{45}{6} =$ $\frac{32}{12} =$ $\frac{42}{9} =$

4. $\frac{36}{12} =$ $\frac{28}{10} =$ $\frac{11}{2} =$ $\frac{44}{16} =$ $\frac{50}{4} =$

**Write the next division problems as improper fractions. Then change the improper fractions to mixed numbers. Reduce any remaining fractions.**

5. It took Mrs. Pagan 100 minutes to drive to the nearest shopping center. Change the time to hours. (1 hour = 60 minutes)

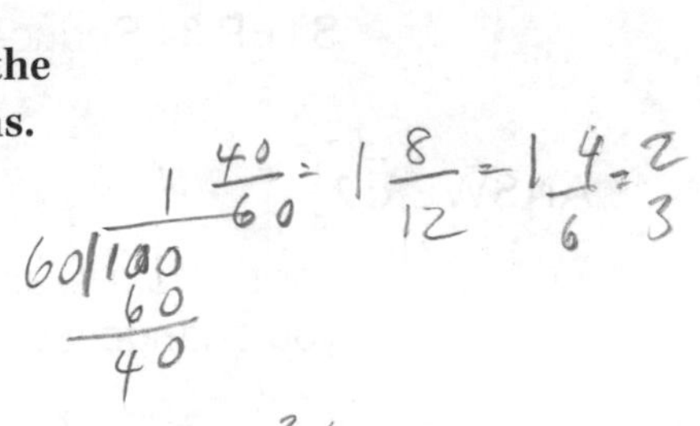

6. Steve cut a board 8 feet long into three equal pieces. How long was each piece?

7. An elevator can carry a load of 2,500 pounds. How many tons can it carry? (1 ton = 2,000 pounds)

8. In a school where adult education classes are held, each classroom can hold 30 students. How many classrooms are needed for 255 students?

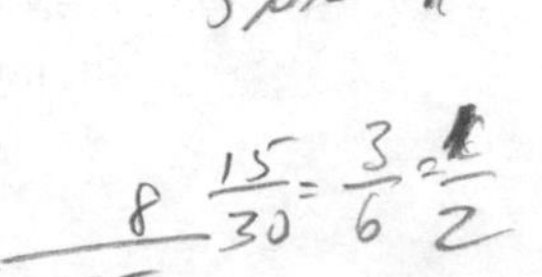

# Changing Mixed Numbers to Improper Fractions

An important skill in multiplication and division of fractions is changing a mixed number, such as $2\frac{1}{4}$, to an improper fraction. You know that one whole equals $\frac{4}{4}$ and that 2 equals $\frac{8}{4}$. Adding the extra $\frac{1}{4}$, you get $\frac{9}{4}$.

$\frac{4}{4} + \frac{4}{4} + \frac{1}{4} = \frac{9}{4}$

Study the next examples to see how to change mixed numbers to improper fractions.

**EXAMPLE 1** **Change $2\frac{1}{4}$ to an improper fraction.**

**STEP 1** Multiply the denominator, 4, by the whole number, 2. $4 \times 2 = 8$

**STEP 2** Add the numerator, 1, to 8. $8 + 1 = 9$

**STEP 3** Write the total, 9, over the denominator, 4. $2\frac{1}{4} = \mathbf{\frac{9}{4}}$

**EXAMPLE 2** **Change $5\frac{2}{3}$ to an improper fraction.**

**STEP 1** Multiply the denominator, 3, by the whole number, 5. $3 \times 5 = 15$

**STEP 2** Add the numerator, 2, to 15. $15 + 2 = 17$

**STEP 3** Write the total, 17, over the denominator, 3. $5\frac{2}{3} = \mathbf{\frac{17}{3}}$

**Change each mixed number to an improper fraction.**

| | | | | |
|---|---|---|---|---|
| **1.** $2\frac{3}{4} =$ | $1\frac{4}{7} =$ | $5\frac{1}{3} =$ | $6\frac{2}{7} =$ | $4\frac{3}{5} =$ |
| **2.** $9\frac{1}{2} =$ | $7\frac{5}{8} =$ | $2\frac{9}{10} =$ | $8\frac{3}{4} =$ | $3\frac{5}{9} =$ |
| **3.** $10\frac{1}{3} =$ | $11\frac{2}{5} =$ | $4\frac{5}{12} =$ | $6\frac{7}{8} =$ | $12\frac{1}{4} =$ |
| **4.** $3\frac{4}{5} =$ | $6\frac{1}{7} =$ | $9\frac{3}{8} =$ | $2\frac{1}{12} =$ | $5\frac{4}{15} =$ |

# Adding Fractions with the Same Denominators

The answer to an addition problem is called the **sum** or **total.** You know that the sum of a quarter and another quarter is two quarters, or $\frac{1}{2}$ dollar.

To add fractions with the same denominators, add the numerators, and put the total over the denominator.

EXAMPLE 1 $\frac{2}{7} + \frac{3}{7} =$

STEP 1 Add the numerators. $2 + 3 = 5$

STEP 2 Put the total, 5, over the denominator, 7.

$$\frac{2}{7} + \frac{3}{7} = \mathbf{\frac{5}{7}}$$

**Add.**

1. $\frac{2}{9} + \frac{3}{9} = \frac{5}{9}$ $\quad \frac{3}{7} + \frac{1}{7} = \frac{4}{7}$ $\quad \frac{4}{8} + \frac{3}{8} = \frac{7}{8}$ $\quad \frac{5}{12} + \frac{2}{12} = \frac{7}{12}$ $\quad \frac{4}{15} + \frac{7}{15} = \frac{11}{15}$

2. $\frac{3}{16} + \frac{1}{16} + \frac{5}{16} = \frac{9}{16}$ $\quad \frac{5}{9} + \frac{2}{9} + \frac{1}{9} = \frac{8}{9}$ $\quad \frac{2}{15} + \frac{7}{15} + \frac{4}{15} = \frac{13}{15}$ $\quad \frac{9}{20} + \frac{3}{20} + \frac{7}{20} = \frac{19}{20}$ $\quad \frac{1}{24} + \frac{9}{24} + \frac{7}{24} = \frac{17}{24}$

**With mixed numbers, add fractions and whole numbers separately.**

3. $4\frac{2}{5} + 3\frac{1}{5} = 8\frac{3}{5}$ $\quad 6\frac{3}{10} + 8\frac{6}{10} = 14\frac{9}{10}$ $\quad 5\frac{4}{16} + 4\frac{5}{16} = 9\frac{9}{16}$ $\quad 8\frac{2}{15} + 6\frac{9}{15} = 14\frac{11}{15}$

4. $3\frac{2}{9} + 5\frac{1}{9} + 4\frac{5}{9} = 12\frac{8}{9}$ $\quad 6\frac{7}{24} + 9\frac{3}{24} + 2\frac{9}{24} = 17\frac{19}{24}$ $\quad 7\frac{2}{7} + 8\frac{2}{7} + 5\frac{2}{7} = 20\frac{6}{7}$ $\quad 9\frac{3}{10} + 2\frac{5}{10} + 4\frac{1}{10} = 15\frac{9}{10}$

Sometimes the total of an addition problem can be reduced.

**EXAMPLE 2** $\frac{5}{12} + \frac{1}{12} =$

**STEP 1** Add the numerators. $5 + 1 = 6$

**STEP 2** Place the total over the denominator. $\frac{6}{12}$

**STEP 3** Reduce the answer. $\frac{6 \div 6}{12 \div 6} = \mathbf{\frac{1}{2}}$

$$\begin{array}{r} \frac{5}{12} \\ + \frac{1}{12} \\ \hline \frac{6}{12} = \mathbf{\frac{1}{2}} \end{array}$$

**Add and reduce.**

**5.** $\frac{5}{8} + \frac{1}{8} = \frac{6}{8} = \frac{3}{4}$ $\quad$ $\frac{4}{9} + \frac{2}{9} = \frac{6}{9} = \frac{2}{3}$ $\quad$ $\frac{3}{12} + \frac{5}{12} = \frac{8}{12} = \frac{2}{3}$ $\quad$ $\frac{7}{15} + \frac{3}{15} = \frac{10}{15} = \frac{2}{3}$ $\quad$ $\frac{3}{16} + \frac{5}{16} = \frac{8}{16} = \frac{1}{2}$

**6.** $\frac{2}{10} + \frac{3}{10} + \frac{3}{10} = \frac{8}{10} = \frac{4}{5}$ $\quad$ $\frac{6}{15} + \frac{2}{15} + \frac{4}{15} = \frac{12}{15} = \frac{4}{5}$ $\quad$ $\frac{7}{20} + \frac{3}{20} + \frac{4}{20} = \frac{14}{20} = \frac{7}{10}$ $\quad$ $\frac{9}{24} + \frac{7}{24} + \frac{2}{24} = \frac{18}{24} = \frac{3}{4}$ $\quad$ $\frac{5}{18} + \frac{7}{18} + \frac{2}{18} = \frac{14}{18} = \frac{7}{9}$

**7.** $5\frac{2}{6} + 6\frac{1}{6} = 11\frac{3}{6} = 11\frac{1}{2}$ $\quad$ $7\frac{5}{12} + 9\frac{3}{12} = 16\frac{8}{12} = \frac{2}{3}$ $\quad$ $10\frac{5}{14} + 8\frac{7}{14} = 18\frac{12}{14} = \frac{6}{7}$ $\quad$ $13\frac{8}{15} + 29\frac{2}{15} = 42\frac{10}{15} = \frac{2}{3}$

**8.** $9\frac{2}{9} + 8\frac{1}{9} + 7\frac{3}{9} = 24\frac{6}{9} = \frac{2}{3}$ $\quad$ $6\frac{3}{10} + 7\frac{2}{10} + 4\frac{3}{10} = 17\frac{8}{10} = 17\frac{4}{5}$ $\quad$ $10\frac{5}{16} + 4\frac{3}{16} + 3\frac{4}{16} = 17\frac{12}{16} = 17\frac{3}{4}$ $\quad$ $12\frac{3}{8} + 9\frac{1}{8} + 10\frac{2}{8} = 31\frac{6}{8} = 31\frac{3}{4}$

If the total of an addition problem is an improper fraction, change the improper fraction to a mixed number. (See page 19.)

**EXAMPLE 3** $\frac{5}{8} + \frac{7}{8} =$

**STEP 1** Add the fractions. $\frac{5}{8} + \frac{7}{8} = \frac{12}{8}$.

**STEP 2** Change $\frac{12}{8}$ to a mixed number. $\frac{12}{8} = 1\frac{4}{8}$

**STEP 3** Reduce $1\frac{4}{8}$. $1\frac{4 \div 4}{8 \div 4} = 1\frac{1}{2}$

(In the last step, you could reduce $\frac{12}{8}$ first. $\frac{12}{8} = \frac{3}{2} = \mathbf{1\frac{1}{2}}$)

$$\begin{array}{r} \frac{5}{8} \\ +\frac{7}{8} \\ \hline \frac{12}{8} = 1\frac{4}{8} = \mathbf{1\frac{1}{2}} \end{array}$$

**Add and reduce.**

**9.** $\frac{4}{5} + \frac{3}{5}$ $\qquad$ $\frac{6}{8} + \frac{5}{8}$ $\qquad$ $\frac{7}{10} + \frac{6}{10}$ $\qquad$ $\frac{8}{9} + \frac{5}{9}$ $\qquad$ $\frac{3}{6} + \frac{3}{6}$

**10.** $\frac{11}{12} + \frac{5}{12}$ $\qquad$ $\frac{9}{14} + \frac{7}{14}$ $\qquad$ $\frac{7}{8} + \frac{5}{8}$ $\qquad$ $\frac{11}{15} + \frac{7}{15}$ $\qquad$ $\frac{9}{16} + \frac{15}{16}$

**11.** $\frac{5}{9} + \frac{7}{9} + \frac{8}{9}$ $\qquad$ $\frac{6}{7} + \frac{5}{7} + \frac{3}{7}$ $\qquad$ $\frac{7}{10} + \frac{3}{10} + \frac{5}{10}$ $\qquad$ $\frac{9}{12} + \frac{11}{12} + \frac{7}{12}$ $\qquad$ $\frac{6}{8} + \frac{5}{8} + \frac{7}{8}$

**12.** $8\frac{3}{8} + 3\frac{7}{8} + 5\frac{5}{8}$ $\qquad$ $9\frac{5}{6} + 2\frac{1}{6} + 6\frac{4}{6}$ $\qquad$ $5\frac{7}{10} + 4\frac{9}{10} + 7\frac{5}{10}$ $\qquad$ $2\frac{7}{12} + 8\frac{8}{12} + 6\frac{9}{12}$

# Adding Fractions with Different Denominators

When the fractions in an addition problem do not have the same denominators, rewrite the problem so that each fraction has the same denominator, called a **common denominator.**

To add a half dollar and a quarter, think of the half dollar as two quarters. As a fraction problem, a half dollar plus a quarter is $\frac{2}{4} + \frac{1}{4} = \frac{3}{4}$ dollar.

A common denominator is a number that can be divided evenly by all the denominators in a problem. The *smallest* number that can be divided evenly by all the denominators in a problem is called the **lowest common denominator** or **LCD.**

Sometimes the largest denominator in a problem is the LCD.

EXAMPLE $\frac{3}{5} + \frac{4}{15} =$

**STEP 1** Since 5 divides evenly into 15, the LCD is 15.

**STEP 2** Raise $\frac{3}{5}$ to $\frac{9}{15}$. (See page 18.)

**STEP 3** Add the new fractions.

$$\frac{3}{5} = \frac{9}{15}$$
$$+\frac{4}{15} = \frac{4}{15}$$
$$\frac{13}{15}$$

**Add and reduce.**

1. $\frac{3}{4} + \frac{1}{2}$ $\quad \frac{2}{3} + \frac{5}{6}$ $\quad \frac{7}{8} + \frac{3}{4}$ $\quad \frac{5}{6} + \frac{1}{3}$ $\quad \frac{5}{9} + \frac{2}{3}$

2. $\frac{3}{8} + \frac{3}{4} + \frac{1}{2}$ $\quad \frac{1}{6} + \frac{5}{12} + \frac{3}{4}$ $\quad \frac{2}{5} + \frac{1}{2} + \frac{9}{10}$ $\quad \frac{2}{3} + \frac{5}{12} + \frac{1}{4}$ $\quad \frac{3}{5} + \frac{1}{3} + \frac{4}{15}$

3. $\frac{3}{4} + \frac{1}{2} + \frac{3}{20}$ $\quad \frac{2}{3} + \frac{5}{6} + \frac{1}{2}$ $\quad \frac{5}{24} + \frac{3}{8} + \frac{1}{3}$ $\quad \frac{4}{9} + \frac{5}{6} + \frac{7}{18}$ $\quad \frac{7}{10} + \frac{1}{3} + \frac{11}{30}$

# Finding a Common Denominator

Here are two ways of finding a common denominator when the largest denominator in an addition problem doesn't work.

1. **Multiply the denominators together.** This method always works, but sometimes it results in a common denominator that is larger than it needs to be.

2. **Go through the multiplication table of the largest denominator.** As you go through the multiplication table of the largest denominator, check each product. When you find the first product that can be divided evenly by the other denominators, you have found the lowest common denominator.

EXAMPLE 1 $\frac{2}{5} + \frac{3}{4} =$

**STEP 1** Multiply the denominators. $5 \times 4 = 20$. 20 is the LCD.

**STEP 2** Raise each fraction to 20ths as on page 18.

**STEP 3** Add the new fractions.

**STEP 4** Change the answer to a mixed number.

$$\frac{2}{5} = \frac{8}{20}$$
$$+\frac{3}{4} = \frac{15}{20}$$
$$\frac{23}{20} = \mathbf{1\frac{3}{20}}$$

EXAMPLE 2 $\frac{2}{3} + \frac{5}{6} + \frac{3}{4} =$

**STEP 1** Go through the multiplication table of the 6's.
$6 \times 1 = 6$, which cannot be divided by 4.
$6 \times 2 = 12$, which can be divided by 3 and 4.

**STEP 2** Raise each fraction to 12ths.

**STEP 3** Add the new fractions.

**STEP 4** Change the answer to a mixed number and reduce.

$$\frac{2}{3} = \frac{8}{12}$$
$$\frac{5}{6} = \frac{10}{12}$$
$$+\frac{3}{4} = \frac{9}{12}$$
$$\frac{27}{12} = 2\frac{3}{12} = \mathbf{2\frac{1}{4}}$$

**Add and reduce.**

**1.** $\frac{3}{5} + \frac{2}{3}$ $\qquad \frac{3}{4} + \frac{1}{3}$ $\qquad \frac{2}{5} + \frac{1}{2}$ $\qquad \frac{3}{7} + \frac{1}{3}$ $\qquad \frac{5}{6} + \frac{2}{5}$

**2.** $\frac{4}{7} + \frac{2}{7}$ $\qquad \frac{5}{6} + \frac{4}{5}$ $\qquad \frac{3}{8} + \frac{5}{8}$ $\qquad \frac{2}{3} + \frac{3}{5}$ $\qquad \frac{5}{9} + \frac{3}{4}$

**3.** $\frac{5}{6} + \frac{3}{4}$ | $\frac{4}{9} + \frac{5}{6}$ | $\frac{7}{10} + \frac{3}{4}$ | $\frac{5}{12} + \frac{5}{9}$ | $\frac{3}{8} + \frac{5}{6}$

**4.** $\frac{2}{3} + \frac{5}{8} + \frac{3}{4}$ | $\frac{1}{4} + \frac{3}{5} + \frac{7}{10}$ | $\frac{8}{9} + \frac{5}{6} + \frac{3}{4}$ | $\frac{5}{16} + \frac{5}{8} + \frac{1}{2}$ | $\frac{2}{9} + \frac{1}{2} + \frac{5}{6}$

**5.** $\frac{5}{6} + \frac{2}{5} + \frac{4}{15}$ | $\frac{7}{12} + \frac{5}{8} + \frac{3}{4}$ | $\frac{2}{3} + \frac{1}{6} + \frac{11}{12}$ | $\frac{7}{20} + \frac{3}{8} + \frac{3}{10}$ | $\frac{4}{9} + \frac{1}{6} + \frac{5}{12}$

**6.** $\frac{2}{3} + \frac{4}{9} + \frac{5}{6}$ | $\frac{6}{7} + \frac{3}{4} + \frac{1}{2}$ | $\frac{2}{9} + \frac{1}{3} + \frac{1}{6}$ | $\frac{5}{7} + \frac{4}{9} + \frac{2}{3}$ | $\frac{11}{16} + \frac{1}{3} + \frac{7}{8}$

**Rewrite each problem with whole numbers under whole numbers and fractions under fractions. Then add and reduce.**

**7.** $4\frac{3}{5} + 6\frac{3}{4} =$ $\qquad$ $7\frac{5}{8} + 9\frac{2}{3} =$ $\qquad$ $8\frac{5}{9} + 3\frac{2}{3} =$

**8.** $6\frac{5}{12} + 7\frac{3}{8} =$ $\qquad$ $10\frac{2}{7} + 8\frac{1}{3} =$ $\qquad$ $6\frac{1}{6} + 4\frac{3}{4} =$

**9.** $7\frac{9}{10} + 8\frac{1}{4} =$ $3\frac{4}{9} + 12\frac{5}{6} =$ $9\frac{5}{8} + 3\frac{7}{12} =$

**10.** $8\frac{5}{6} + 2\frac{2}{9} =$ $3\frac{7}{8} + 11\frac{3}{5} =$ $6\frac{9}{10} + 5\frac{2}{3} =$

**11.** $5\frac{2}{3} + 9\frac{3}{5} + 2\frac{7}{10} =$ $3\frac{3}{8} + 8\frac{1}{6} + 7\frac{3}{4} =$ $2\frac{2}{9} + 10\frac{3}{4} + 4\frac{5}{6} =$

**12.** $6\frac{1}{2} + 1\frac{1}{3} + 14\frac{1}{4} =$ $7\frac{7}{8} + \frac{2}{3} + 4\frac{1}{2} =$ $9\frac{5}{12} + 11\frac{2}{3} + \frac{4}{9} =$

**13.** $6\frac{7}{16} + \frac{1}{2} + 8\frac{5}{8} =$ $12\frac{3}{4} + 9\frac{5}{6} + \frac{2}{3} =$ $9\frac{1}{2} + 3\frac{3}{4} + 2\frac{2}{3} =$

**14.** $\frac{6}{7} + 15\frac{3}{5} + 7\frac{7}{10} =$ $4\frac{5}{12} + \frac{4}{9} + 19\frac{3}{4} =$ $11\frac{7}{8} + 5\frac{3}{7} + \frac{1}{2} =$

# Estimating Addition Answers

So far in this book, you have found exact answers. Sometimes an **approximate** answer or an **estimate** is all you need. For example, an estimate can tell you whether an answer is reasonable. Review "Thinking About the Size of Fractions" on page 14. When you add two fractions, think about the sum. Is the sum equal to one whole (= 1), less than one whole (< 1), or more than one whole (> 1)? Study these examples carefully.

EXAMPLES $\frac{1}{2} + \frac{1}{2}$, the sum is 1. $\frac{1}{2} + \frac{3}{6} = 1$

$\frac{1}{2}$ + a fraction that is less than $\frac{1}{2}$ = a number less than 1. $\frac{1}{2} + \frac{2}{5} < 1$

$\frac{1}{2}$ + a fraction that is more than $\frac{1}{2}$ = a number more than 1. $\frac{1}{2} + \frac{4}{5} > 1$

**In each box, write <, >, or = to make the statement true.**

1. $\frac{5}{16} + \frac{1}{2}$ ☐ 1 $\frac{4}{8} + \frac{1}{2}$ ☐ 1 $\frac{1}{2} + \frac{9}{10}$ ☐ 1 $\frac{3}{6} + \frac{2}{4}$ ☐ 1

2. $\frac{1}{2} + \frac{3}{8}$ ☐ 1 $\frac{2}{3} + \frac{5}{6}$ ☐ 1 $\frac{9}{10} + \frac{2}{3}$ ☐ 1 $\frac{3}{8} + \frac{4}{9}$ ☐ 1

To estimate the answer to an addition problem with mixed numbers, first round each mixed number to the nearest whole number. For a mixed number with a fraction of $\frac{1}{2}$ or more, round to the next whole number.

EXAMPLES $8\frac{2}{3} \rightarrow 9$ $1\frac{7}{12} \rightarrow 2$ $4\frac{1}{2} \rightarrow 5$

For a mixed number with a fraction less than $\frac{1}{2}$, drop the fraction and use the whole number.

EXAMPLES $7\frac{3}{8} \rightarrow 7$ $10\frac{1}{3} \rightarrow 10$ $1\frac{2}{5} \rightarrow 1$

EXAMPLE Solve the problem $7\frac{1}{16} + 4\frac{3}{4}$. Then round each mixed number to the nearest whole number, and add the rounded numbers.

Exact answer: $7\frac{1}{16} + 4\frac{3}{4} = 7\frac{1}{16} + 4\frac{12}{16} = \mathbf{11\frac{13}{16}}$

Estimate: $7 + 5 = \mathbf{12}$

The symbol ≈ means "approximately equal to." For the last example, $7\frac{1}{16} + 4\frac{3}{4} \approx 12$.

**For each problem, first round each mixed number to the nearest whole number. Then add the rounded numbers.**

3. $1\frac{1}{2} + 2\frac{3}{4} \approx$ $4\frac{1}{8} + 5\frac{1}{3} \approx$ $3\frac{2}{3} + 7\frac{5}{6} \approx$

4. $2\frac{3}{10} + 4\frac{1}{2} \approx$ $4\frac{7}{8} + 9\frac{3}{10} \approx$ $2\frac{1}{6} + 5\frac{1}{2} \approx$

5. $5\frac{1}{3} + 1\frac{3}{10} \approx$ $9\frac{1}{2} + 6\frac{5}{7} \approx$ $8\frac{2}{5} + 4\frac{3}{4} \approx$

**For more practice, find the exact answers for the problems above.**

# Applying Your Addition Skills

An addition problem may ask you to *combine* measurements or to find a *total.* Use estimation to find out whether an answer is reasonable.

**Solve and write the correct label, such as inches or miles. Then round each fraction or mixed number to the nearest whole number and add the rounded numbers to estimate the answer.**

**1.** Karen is $62\frac{1}{2}$ inches tall. Her mother is $5\frac{3}{4}$ inches taller. How tall is Karen's mother?

**2.** Doing errands on Monday, Mrs. Johnson drove $5\frac{1}{2}$ miles to the supermarket, $3\frac{7}{10}$ miles to the hardware store, $\frac{8}{10}$ mile to the laundromat, and $6\frac{1}{10}$ miles back home. What total distance did she drive?

**3.** Mr. Munro's empty suitcase weighs $4\frac{3}{4}$ pounds. The items he packed in his suitcase weigh $17\frac{3}{5}$ pounds. What was the weight of his suitcase when he filled it?

**4.** When Petra went shopping, she bought 2 pounds of sugar, $3\frac{1}{4}$ pounds of ground beef, $2\frac{2}{3}$ pounds of cheese, and a $\frac{7}{8}$-pound can of tomatoes. What was the total weight of her purchases?

**5.** John does carpentry part-time. One week he spent his evenings converting his neighbor's attic into an extra bedroom. Monday night he worked $3\frac{1}{2}$ hours. Tuesday night he worked $4\frac{1}{3}$ hours. Wednesday he worked $2\frac{3}{4}$ hours, and Thursday he worked $3\frac{2}{3}$ hours. How many hours did John work on the attic that week?

**6.** When Ruby was sick, her weight went down to $116\frac{1}{5}$ pounds. By the time she recovered, she had gained $12\frac{1}{2}$ pounds. What was Ruby's final weight when she was well?

**7.** Lois talked on the phone for $\frac{1}{2}$ hour this morning, $\frac{3}{5}$ hour in the afternoon, and $1\frac{2}{3}$ hours in the evening. How much time did Lois spend on the phone today?

# Subtracting Fractions with the Same Denominators

The answer to a subtraction problem is called the **difference.** To subtract fractions with the same denominators, subtract the numerators and put the difference over the denominator.

EXAMPLE $\frac{5}{12} - \frac{1}{12} =$

**STEP 1** Subtract the numerators. $5 - 1 = 4$

**STEP 2** Put the difference, 4, over the denominator. $\frac{4}{12}$

**STEP 3** Reduce $\frac{4}{12}$ by dividing 4 and 12 by 4.

$$\frac{5}{12} - \frac{1}{12} = \frac{4}{12} = \mathbf{\frac{1}{3}}$$

**Subtract and reduce.**

**1.** $\frac{5}{9} - \frac{2}{9}$ $\qquad \frac{7}{10} - \frac{6}{10}$ $\qquad \frac{5}{8} - \frac{1}{8}$ $\qquad \frac{7}{12} - \frac{5}{12}$ $\qquad \frac{9}{20} - \frac{3}{20}$

**2.** $\frac{13}{15} - \frac{8}{15}$ $\qquad \frac{15}{16} - \frac{9}{16}$ $\qquad \frac{23}{24} - \frac{11}{24}$ $\qquad \frac{11}{18} - \frac{5}{18}$ $\qquad \frac{17}{20} - \frac{13}{20}$

**With mixed numbers, subtract fractions and whole numbers separately.**

**3.** $8\frac{6}{7} - 5\frac{2}{7}$ $\qquad 10\frac{5}{8} - 4\frac{5}{8}$ $\qquad 7\frac{8}{9} - 6\frac{5}{9}$ $\qquad 13\frac{9}{10} - 9\frac{5}{10}$

**4.** $15\frac{7}{16} - 7\frac{3}{16}$ $\qquad 14\frac{11}{12} - 5\frac{5}{12}$ $\qquad 18\frac{8}{15} - 9\frac{2}{15}$ $\qquad 23\frac{5}{6} - 7\frac{1}{6}$

# Subtracting Fractions with Different Denominators

If the denominators in a subtraction problem are different, find the LCD and raise the fractions to higher terms. Then follow the rules on page 31.

EXAMPLE $\frac{5}{8} - \frac{1}{3} =$

**STEP 1** The LCD is $8 \times 3 = 24$. $\quad \frac{5}{8} = \frac{15}{24}$

**STEP 2** Raise each fraction to 24ths. $\quad -\frac{1}{3} = \frac{8}{24}$

**STEP 3** Subtract the new fractions. $\quad \mathbf{\frac{7}{24}}$

**Subtract and reduce.**

| | | | | | |
|---|---|---|---|---|---|
| **1.** | $\frac{3}{4}$<br>$-\frac{1}{2}$ | $\frac{5}{8}$<br>$-\frac{1}{4}$ | $\frac{5}{6}$<br>$-\frac{1}{3}$ | $\frac{3}{4}$<br>$-\frac{3}{16}$ | $\frac{1}{2}$<br>$-\frac{3}{10}$ |
| **2.** | $\frac{2}{3}$<br>$-\frac{1}{4}$ | $\frac{4}{5}$<br>$-\frac{1}{3}$ | $\frac{7}{8}$<br>$-\frac{2}{3}$ | $\frac{5}{6}$<br>$-\frac{3}{5}$ | $\frac{5}{9}$<br>$-\frac{1}{6}$ |
| **3.** | $8\frac{11}{12}$<br>$2\frac{3}{8}$ | $9\frac{5}{7}$<br>$-3\frac{1}{2}$ | $12\frac{4}{5}$<br>$-5\frac{2}{9}$ | $11\frac{3}{4}$<br>$-8\frac{7}{10}$ | |
| **4.** | $13\frac{4}{7}$<br>$-7\frac{3}{8}$ | $21\frac{8}{9}$<br>$-6\frac{1}{4}$ | $9\frac{5}{6}$<br>$-2\frac{2}{5}$ | $20\frac{11}{12}$<br>$-8\frac{8}{12}$ | |

**Rewrite each problem with the first number on top and the second number below it. Then subtract and reduce.**

**5.** $7\frac{1}{2} - 2\frac{1}{4} =$ $12\frac{3}{4} - 9\frac{3}{8} =$ $5\frac{2}{3} - 3\frac{3}{5} =$

**6.** $4\frac{11}{16} - 2\frac{3}{8} =$ $18\frac{7}{10} - 9\frac{1}{2} =$ $22\frac{8}{15} - 17\frac{2}{9} =$

**7.** $8\frac{9}{28} - 7\frac{2}{7} =$ $35\frac{11}{30} - 28\frac{1}{3} =$ $14\frac{11}{12} - 7\frac{5}{8} =$

**8.** $9\frac{11}{36} - 5\frac{2}{9} =$ $15\frac{7}{8} - 8\frac{5}{12} =$ $45\frac{8}{9} - 38\frac{5}{6} =$

**9.** $5\frac{7}{8} - 4\frac{5}{9} =$ $10\frac{5}{6} - 6\frac{1}{4} =$ $100\frac{5}{8} - 98\frac{1}{6} =$

**10.** $11\frac{1}{3} - 5\frac{1}{8} =$ $24\frac{5}{6} - 21\frac{4}{5} =$ $8\frac{3}{4} - 6\frac{2}{5} =$

**11.** $17\frac{3}{4} - 9\frac{2}{3} =$ $15\frac{1}{3} - 7\frac{3}{10} =$ $12\frac{1}{2} - 8\frac{4}{9} =$

**12.** $13\frac{5}{8} - 11\frac{5}{12} =$ $16\frac{5}{6} - 9\frac{7}{10} =$ $18\frac{3}{5} - 9\frac{3}{10} =$

**13.** $11\frac{2}{3} - 5\frac{1}{2} =$ $7\frac{1}{2} - 3\frac{2}{5} =$ $19\frac{7}{12} - 8\frac{3}{8} =$

**14.** $21\frac{14}{15} - 18\frac{7}{10} =$ $13\frac{11}{18} - 8\frac{1}{2} =$ $25\frac{5}{8} - 22\frac{2}{5} =$

# Borrowing and Subtracting Fractions

In order to have a fraction to subtract from, you sometimes have to borrow from a whole number. Look at the examples carefully.

EXAMPLE 1 $9 - 6\frac{3}{5} =$

Since there is nothing to subtract the $\frac{3}{5}$ from, you have to borrow.

**STEP 1** Borrow 1 from the 9 and change the 1 to 5ths because 5 is the LCD.

$1 = \frac{5}{5}$

**STEP 2** Subtract the fractions and the whole numbers.

$$\begin{array}{rcl} 9 & = & 8\frac{5}{5} \\ -6\frac{3}{5} & = & 6\frac{3}{5} \\ \hline & & \mathbf{2\frac{2}{5}} \end{array}$$

EXAMPLE 2 $12\frac{3}{7} - 8\frac{6}{7} =$

Since $\frac{6}{7}$ is larger than $\frac{3}{7}$, you have to borrow.

**STEP 1** Borrow 1 from 12 and change the 1 to 7ths because 7 is the LCD.

$1 = \frac{7}{7}$

$$\begin{array}{rcl} 12\frac{3}{7} & = & 11\frac{7}{7} + \frac{3}{7} \\ -\ 8\frac{6}{7} & & \\ \hline \end{array}$$

**STEP 2** Add $\frac{7}{7}$ to $\frac{3}{7}$. $\frac{7}{7} + \frac{3}{7} = \frac{10}{7}$

**STEP 3** Subtract the fractions and the whole numbers.

$$\begin{array}{rcl} 11\frac{7}{7} + \frac{3}{7} & = & 11\frac{10}{7} \\ -\ 8\frac{6}{7} & = & 8\frac{6}{7} \\ \hline & & \mathbf{3\frac{4}{7}} \end{array}$$

EXAMPLE 3 $8\frac{1}{3} - 4\frac{3}{4} =$

**STEP 1** Raise each fraction to 12ths because 12 is the LCD.

$$\begin{array}{rcl} 8\frac{1}{3} & = & 8\frac{4}{12} \\ -4\frac{3}{4} & = & 4\frac{9}{12} \\ \hline \end{array}$$

**STEP 2** Borrow 1 from 8 and change the 1 to 12ths.

$1 = \frac{12}{12}$

$$\begin{array}{rcl} 8\frac{4}{12} & = & 7\frac{12}{12} + \frac{4}{12} \\ -4\frac{9}{12} & & \\ \hline \end{array}$$

**STEP 3** Add $\frac{12}{12}$ to $\frac{4}{12}$. $\frac{12}{12} + \frac{4}{12} = \frac{16}{12}$

**STEP 4** Subtract the fractions and the whole numbers.

$$\begin{array}{rcl} 7\frac{12}{12} + \frac{4}{12} & = & 7\frac{16}{12} \\ -4\frac{9}{12} & = & 4\frac{9}{12} \\ \hline & & \mathbf{3\frac{7}{12}} \end{array}$$

**Subtract and reduce.**

1. $8 - \frac{5}{6}$ $\qquad$ $4 - \frac{3}{7}$ $\qquad$ $12 - \frac{1}{2}$ $\qquad$ $9 - \frac{2}{5}$ $\qquad$ $10 - \frac{8}{11}$

2. $12 - 8\frac{3}{7}$ $\qquad$ $9 - 5\frac{2}{3}$ $\qquad$ $7 - 5\frac{7}{12}$ $\qquad$ $6 - 2\frac{5}{9}$ $\qquad$ $10 - 3\frac{5}{16}$

3. $8\frac{2}{9} - 4\frac{5}{9}$ $\qquad$ $11\frac{3}{8} - 4\frac{7}{8}$ $\qquad$ $14\frac{7}{12} - 6\frac{11}{12}$ $\qquad$ $15\frac{1}{5} - 8\frac{4}{5}$

4. $12\frac{7}{15} - 7\frac{8}{15}$ $\qquad$ $22\frac{4}{7} - 6\frac{6}{7}$ $\qquad$ $19\frac{1}{3} - 12\frac{2}{3}$ $\qquad$ $18\frac{3}{16} - 10\frac{7}{16}$

5. $20\frac{1}{8} - 9\frac{5}{8}$ $\qquad$ $36\frac{3}{10} - 8\frac{9}{10}$ $\qquad$ $15\frac{7}{20} - 14\frac{13}{20}$ $\qquad$ $30\frac{5}{9} - 27\frac{8}{9}$

6. $7\frac{1}{3} - 4\frac{11}{12}$ $\qquad$ $15\frac{1}{6} - 3\frac{2}{3}$ $\qquad$ $10\frac{2}{5} - 5\frac{7}{10}$ $\qquad$ $5\frac{2}{3} - 2\frac{3}{4}$

**Rewrite each problem with the first number on top and the second number below it. Then subtract and reduce.**

**7.** $12\frac{2}{5} - 6\frac{3}{4} =$ $\qquad$ $14\frac{3}{8} - 9\frac{3}{4} =$ $\qquad$ $7\frac{2}{3} - 2\frac{8}{9} =$

**8.** $18\frac{1}{6} - 3\frac{3}{4} =$ $\qquad$ $25\frac{1}{2} - 6\frac{4}{7} =$ $\qquad$ $20\frac{5}{12} - 11\frac{5}{6} =$

**9.** $19\frac{1}{2} - 13\frac{7}{10} =$ $\qquad$ $36\frac{4}{9} - 4\frac{3}{5} =$ $\qquad$ $17\frac{1}{3} - 8\frac{5}{6} =$

**10.** $15\frac{3}{8} - 7\frac{4}{5} =$ $\qquad$ $30\frac{1}{3} - 16\frac{7}{10} =$ $\qquad$ $12\frac{1}{6} - 10\frac{7}{12} =$

**11.** $14\frac{1}{2} - 13\frac{11}{15} =$ $\qquad$ $18\frac{4}{9} - 14\frac{3}{4} =$ $\qquad$ $22\frac{5}{8} - 8\frac{4}{5} =$

**12.** $16\frac{7}{12} - 9\frac{7}{8} =$ $\qquad$ $28\frac{1}{6} - 17\frac{3}{5} =$ $\qquad$ $15\frac{3}{4} - 8\frac{7}{8} =$

**13.** $30\frac{1}{4} - 16\frac{5}{12}$ $\qquad$ $19\frac{3}{8} - 18\frac{1}{2} =$ $\qquad$ $17\frac{2}{7} - 15\frac{5}{8} =$

**14.** $35\frac{1}{4} - 18\frac{3}{5} =$ $\qquad$ $13\frac{2}{9} - 7\frac{5}{6} =$ $\qquad$ $24\frac{3}{16} - 9\frac{2}{3} =$

# Estimating Subtraction Answers

To estimate the answer to a subtraction problem with mixed numbers, first round each mixed number to the nearest whole number. Then subtract the rounded numbers. Compare these two similar examples.

EXAMPLE Solve the problem $8 - 4\frac{5}{6}$. Then estimate the answer.

Exact answer: $8 - 4\frac{5}{6} = 7\frac{6}{6} - 4\frac{5}{6} = \mathbf{3\frac{1}{6}}$

Estimate: $8 - 5 = \mathbf{3}$

EXAMPLE Solve the problem $8 - 4\frac{1}{6}$. Then estimate the answer.

Exact answer: $8 - 4\frac{1}{6} = 7\frac{6}{6} - 4\frac{1}{6} = \mathbf{3\frac{5}{6}}$

Estimate: $8 - 4 = \mathbf{4}$

**For problems 1 to 5, choose the best whole number problem to estimate each answer.**

1. $5\frac{3}{4} - 2\frac{1}{8} \approx$ **a.** $6 - 3 =$ **b.** $5 - 2 =$ **c.** $6 - 2 =$

2. $10\frac{5}{12} - 9\frac{1}{3} \approx$ **a.** $10 - 9 =$ **b.** $11 - 9 =$ **c.** $10 - 10 =$

3. $17 - 2\frac{9}{10} \approx$ **a.** $17 - 4 =$ **b.** $17 - 3 =$ **c.** $17 - 2 =$

4. $6\frac{1}{8} - 4\frac{7}{8} \approx$ **a.** $6 - 4 =$ **b.** $6 - 5 =$ **c.** $7 - 4 =$

5. $2\frac{5}{6} - \frac{3}{16} \approx$ **a.** $3 - 0 =$ **b.** $3 - 1 =$ **c.** $2 - 1 =$

**For each problem, first round each mixed number to the nearest whole number. Then subtract the rounded numbers.**

6. $9\frac{1}{2} - 6\frac{3}{4} \approx$ $5 - 2\frac{3}{8} \approx$ $16\frac{8}{9} - 12 \approx$

7. $8\frac{11}{12} - 1\frac{1}{3} \approx$ $12 - 9\frac{5}{6} \approx$ $10\frac{1}{4} - 5\frac{11}{12} \approx$

8. $1\frac{15}{16} - 1\frac{1}{4} \approx$ $11\frac{1}{3} - 4\frac{4}{5} \approx$ $7\frac{1}{6} - 2\frac{9}{10} \approx$

**For more practice, find the exact answers for the problems above.**

# Applying Your Subtraction Skills

A subtraction problem may ask you to figure out what is *left* after you take something away or to find how much something *increases* or *decreases.*

**For problems 1 to 8, solve and write the correct label, such as inches or pounds, next to each answer. Reduce each answer to lowest terms. Then use rounding to estimate each answer.**

1. From a board $38\frac{1}{2}$ inches long, Pete cut a piece $17\frac{5}{8}$ inches long. How long was the remaining piece?

2. Jeff weighed 166 pounds. He went on a diet and lost $11\frac{3}{4}$ pounds. How much did Jeff weigh after his diet?

3. Before leaving on a weekend trip, Mr. Green noticed that his mileage gauge registered $20{,}245\frac{3}{10}$ miles. When he returned home, the gauge registered $20{,}734\frac{7}{10}$ miles. How many miles did Mr. Green drive that weekend?

4. Adrienne works as a seamstress. From a piece of cloth 5 yards long, she used $1\frac{2}{3}$ yards to make a new curtain for her bathroom. How long was the remaining piece of cloth?

5. From a 10-foot long pipe, Shirley cut a section $1\frac{11}{12}$ feet long to repair her kitchen drain. How long was the piece of pipe that was left?

6. Tom changed the rotating speed of his cement mixer from $6\frac{1}{3}$ rpm's to 10 rpm's. By how many revolutions per minute did the speed of the mixer increase?

7. From a 100-pound bag of cement, Fred used $44\frac{5}{8}$ pounds to make concrete. How much cement was left in the bag?

8. Esther bought a $1\frac{3}{4}$-pound bar of baking chocolate. If she used $\frac{5}{8}$ pound of chocolate to make a cake, how much chocolate was left?

**Read each of the following problems carefully to decide whether to add or to subtract.**

Mr. Alonso bought 60 feet of nylon rope. To tie a mattress to the roof of his car, he used $18\frac{2}{3}$ feet of the rope. Then to pitch a tent for his son, he used $24\frac{1}{2}$ feet of rope.

**9.** How much rope did Mr. Alonso use to tie the mattress to the roof of his car and to pitch the tent?

**10.** After Mr. Alonso used the two pieces of rope, how much rope was left from the original 60 feet?

Mary ran $2\frac{3}{4}$ miles on Monday, $3\frac{3}{8}$ miles on Wednesday, and $1\frac{5}{8}$ miles on Friday.

**11.** How many miles did Mary run in the three days?

**12.** Mary tries to run 10 miles each week. How many more miles does Mary need to run to complete 10 miles?

A town needs $3 million for a new recreation center. So far the town has received $$\frac{3}{4}$ million from private gifts, $$1\frac{1}{8}$ million from a state grant, and $$\frac{1}{2}$ million from a federal grant.

**13.** What total amount has the town received so far?

**14.** How much more money does the town need to reach its goal?

Selma works for an airline. One of her duties is to check the size of each passenger's carry-on luggage. The airline allows carry-on bags with a combined length plus width plus height of no more than 37 inches.

**15.** Mrs. Burke's bag is $15\frac{3}{4}$ inches long, $9\frac{1}{2}$ inches wide, and $12\frac{7}{8}$ inches high. Does her bag fit within the guidelines for carry-on bags?

**16.** Mr. Burke's bag is $16\frac{1}{2}$ inches long, $8\frac{5}{8}$ inches wide, and $11\frac{3}{8}$ inches high. Does his bag fit within the airline's guidelines for carry-on bags?

# Multiplying Fractions

The answer to a multiplication problem is called the **product.** Multiplying fractions means *finding a fraction of a fraction.* When you multiply proper fractions, the product is *smaller* than the fractions you multiply. You are finding a *part of a part.*

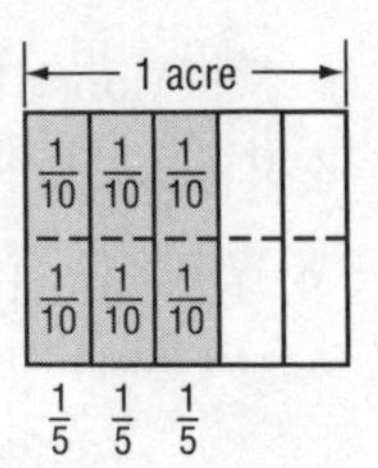


To multiply fractions, multiply the numerators together and multiply the denominators together.

EXAMPLE The owner of a $\frac{3}{5}$-acre parcel of land plans to sell $\frac{1}{2}$ of the land. How large is the parcel that he plans to sell?

Find $\frac{1}{2}$ of $\frac{3}{5}$.

STEP 1 Multiply the numerators. $1 \times 3 = 3$

STEP 2 Multiply the denominators. $2 \times 5 = 10$

$$\frac{1}{2} \times \frac{3}{5} = \mathbf{\frac{3}{10}}$$

ANSWER: The size of the parcel of land is $\mathbf{\frac{3}{10}}$ **acre.**

**Multiply and reduce.**

1. $\frac{2}{3} \times \frac{4}{5} = \frac{8}{15}$ $\quad \frac{5}{7} \times \frac{2}{9} = \frac{10}{63}$ $\quad \frac{1}{8} \times \frac{7}{10} = \frac{7}{80}$ $\quad \frac{3}{4} \times \frac{5}{8} = \frac{15}{32}$

2. $\frac{1}{3} \times \frac{1}{5} = \frac{1}{15}$ $\quad \frac{4}{7} \times \frac{4}{9} = \frac{16}{63}$ $\quad \frac{5}{6} \times \frac{5}{8} = \frac{25}{48}$ $\quad \frac{9}{10} \times \frac{1}{4} = \frac{9}{40}$

3. $\frac{7}{9} \times \frac{2}{5} = \frac{14}{35}$ $\quad \frac{3}{8} \times \frac{7}{8} = \frac{21}{64}$ $\quad \frac{1}{6} \times \frac{5}{6} = \frac{5}{36}$ $\quad \frac{8}{9} \times \frac{2}{9} = \frac{16}{81}$

**With three fractions, multiply the numerators of the first two fractions together. Then multiply that answer by the third numerator. Do the same with the denominators.**

4. $\frac{3}{5} \times \frac{1}{2} \times \frac{3}{4} = \frac{9}{40}$ $\quad \frac{5}{7} \times \frac{1}{3} \times \frac{1}{2} = \frac{5}{42}$ $\quad \frac{2}{3} \times \frac{1}{3} \times \frac{5}{9} = \frac{10}{81}$

5. $\frac{4}{5} \times \frac{4}{5} \times \frac{1}{3} = \frac{16}{75}$ $\quad \frac{2}{5} \times \frac{7}{9} \times \frac{1}{3} = \frac{14}{135}$ $\quad \frac{1}{3} \times \frac{4}{5} \times \frac{2}{3} = \frac{8}{45}$

# Canceling and Multiplying Fractions

Canceling is a shortcut in the multiplication of fractions. It is just like reducing. It means dividing a numerator and a denominator by a number that goes evenly into both before actually multiplying. You don't have to cancel to get the right answer, but it makes the multiplication easier.

**EXAMPLE** $\frac{10}{21} \times \frac{14}{25} =$

**STEP 1** Cancel 10 and 25 by 5.

$10 \div 5 = 2$ and $25 \div 5 = 5$.

Cross out the 10 and the 25.

$\frac{\cancel{10}^{2}}{21} \times \frac{14}{\cancel{25}_{5}} =$

**STEP 2** Cancel 14 and 21 by 7.

$14 \div 7 = 2$ and $21 \div 7 = 3$.

Cross out the 14 and the 21.

$\frac{2}{\cancel{21}_{3}} \times \frac{\cancel{14}^{2}}{5} = \mathbf{\frac{4}{15}}$

**STEP 3** Multiply across by the new numbers.

$2 \times 2 = 4$ and $3 \times 5 = 15$

---

**Cancel and multiply.**

**1.** $\frac{2}{5} \times \frac{3}{4} =$ $\quad \frac{4}{9} \times \frac{3}{8} =$ $\quad \frac{5}{8} \times \frac{7}{10} =$ $\quad \frac{6}{7} \times \frac{5}{12} =$

**2.** $\frac{4}{5} \times \frac{1}{6} =$ $\quad \frac{8}{15} \times \frac{5}{12} =$ $\quad \frac{9}{20} \times \frac{5}{6} =$ $\quad \frac{4}{5} \times \frac{15}{16} =$

**3.** $\frac{4}{9} \times \frac{3}{8} =$ $\quad \frac{5}{12} \times \frac{9}{10} =$ $\quad \frac{3}{16} \times \frac{8}{9} =$ $\quad \frac{5}{6} \times \frac{9}{10} =$

**4.** $\frac{15}{16} \times \frac{12}{25} =$ $\quad \frac{7}{24} \times \frac{3}{10} =$ $\quad \frac{14}{15} \times \frac{3}{4} =$ $\quad \frac{3}{16} \times \frac{8}{9} =$

**5.** $\frac{9}{16} \times \frac{8}{15} =$ $\quad \frac{6}{7} \times \frac{5}{24} =$ $\quad \frac{5}{6} \times \frac{12}{25} =$ $\quad \frac{8}{15} \times \frac{9}{32} =$

**6.** $\frac{8}{15} \times \frac{3}{16} =$ $\quad \frac{4}{9} \times \frac{9}{14} =$ $\quad \frac{15}{16} \times \frac{4}{5} =$ $\quad \frac{5}{36} \times \frac{9}{20} =$

To multiply more than two fractions, you can cancel when a numerator and a denominator are not next to each other.

**EXAMPLE** $\frac{5}{9} \times \frac{3}{4} \times \frac{16}{25} =$

**STEP 1** Cancel 5 and 25 by 5.

Cancel 3 and 9 by 3.

Cancel 16 and 4 by 4.

$$\frac{\overset{1}{\cancel{5}}}{\underset{3}{\cancel{9}}} \times \frac{\overset{1}{\cancel{3}}}{\underset{1}{\cancel{4}}} \times \frac{\overset{4}{\cancel{16}}}{\underset{5}{\cancel{25}}} = \mathbf{\frac{4}{15}}$$

**STEP 2** Multiply the new numerators.

$1 \times 1 \times 4 = 4$ and $3 \times 1 \times 5 = 15$

**7.** $\frac{7}{8} \times \frac{3}{10} \times \frac{5}{6} =$ $\quad$ $\frac{4}{9} \times \frac{5}{12} \times \frac{3}{5} =$ $\quad$ $\frac{2}{3} \times \frac{14}{15} \times \frac{3}{4} =$

**8.** $\frac{9}{16} \times \frac{5}{8} \times \frac{16}{25} =$ $\quad$ $\frac{4}{15} \times \frac{7}{12} \times \frac{3}{4} =$ $\quad$ $\frac{9}{10} \times \frac{1}{6} \times \frac{5}{8} =$

**9.** $\frac{7}{24} \times \frac{2}{3} \times \frac{4}{5} =$ $\quad$ $\frac{3}{20} \times \frac{18}{25} \times \frac{5}{6} =$ $\quad$ $\frac{11}{12} \times \frac{5}{16} \times \frac{8}{15} =$

**10.** $\frac{4}{5} \times \frac{1}{3} \times \frac{15}{32} =$ $\quad$ $\frac{5}{9} \times \frac{3}{4} \times \frac{4}{15} =$ $\quad$ $\frac{15}{16} \times \frac{7}{10} \times \frac{3}{7} =$

**11.** $\frac{5}{36} \times \frac{18}{25} \times \frac{5}{6} =$ $\quad$ $\frac{9}{10} \times \frac{14}{15} \times \frac{5}{12} =$ $\quad$ $\frac{5}{6} \times \frac{9}{50} \times \frac{4}{9} =$

# Multiplying Fractions and Whole Numbers

Any whole number can be written as a fraction with a denominator of 1. For example, 5 is the same as $\frac{5}{1}$. To check, divide 5 by 1.

**EXAMPLE** $9 \times \frac{5}{6} =$

$$\frac{\overset{3}{\cancel{9}}}{1} \times \frac{5}{\underset{2}{\cancel{6}}} = \frac{15}{2} = \mathbf{7\frac{1}{2}}$$

**STEP 1** Write 9 as a fraction. $9 = \frac{9}{1}$

**STEP 2** Cancel 9 and 6 by 3.

**STEP 3** Multiply across by the new numbers.

**STEP 4** Change the improper fraction to a mixed number (see page 19).

**Multiply and reduce.**

**1.** $4 \times \frac{3}{7} =$   $9 \times \frac{1}{4} =$   $\frac{2}{3} \times 10 =$   $3 \times \frac{4}{5} =$

**2.** $15 \times \frac{2}{3} =$   $\frac{5}{9} \times 18 =$   $\frac{3}{20} \times 12 =$   $\frac{8}{15} \times 45 =$

**3.** $\frac{7}{8} \times 24 =$   $\frac{11}{40} \times 20 =$   $32 \times \frac{7}{16} =$   $12 \times \frac{15}{16} =$

**4.** $35 \times \frac{7}{30} =$   $16 \times \frac{5}{24} =$   $\frac{7}{12} \times 36 =$   $2 \times \frac{9}{10} =$

# Multiplying Mixed Numbers

To multiply with mixed numbers, change every mixed number to an improper fraction. (See page 21.)

EXAMPLE $4\frac{1}{2} \times \frac{5}{6} =$

$\frac{\overset{3}{\cancel{9}}}{2} \times \frac{5}{\underset{2}{\cancel{6}}} = \frac{15}{4} = \mathbf{3\frac{3}{4}}$

**STEP 1** Change $4\frac{1}{2}$ to an improper fraction.
$4\frac{1}{2} = \frac{9}{2}$

**STEP 2** Cancel 9 and 6 by 3.

**STEP 3** Multiply across.

**STEP 4** Change the improper fraction to a mixed number (see page 19).

**Multiply and reduce.**

**1.** $1\frac{1}{2} \times \frac{1}{4} =$ $\quad 1\frac{2}{3} \times \frac{2}{7} =$ $\quad 2\frac{1}{2} \times \frac{7}{8} =$ $\quad \frac{3}{10} \times 5\frac{1}{2} =$

**2.** $\frac{4}{9} \times 3\frac{3}{4} =$ $\quad \frac{2}{7} \times 2\frac{5}{8} =$ $\quad 4\frac{2}{3} \times \frac{15}{16} =$ $\quad 6\frac{3}{7} \times \frac{4}{5} =$

**3.** $2\frac{1}{3} \times 1\frac{1}{5} =$ $\quad 6\frac{2}{3} \times 3\frac{3}{4} =$ $\quad 3\frac{5}{7} \times 4\frac{3}{8} =$ $\quad 16\frac{1}{3} \times 2\frac{5}{14} =$

**4.** $3\frac{3}{4} \times \frac{8}{9} \times 1\frac{1}{5} =$ $\quad 2\frac{2}{5} \times 3\frac{3}{8} \times 2\frac{7}{9} =$ $\quad 2\frac{2}{15} \times 5\frac{1}{4} \times 7\frac{1}{2} =$

# Estimating Multiplication Problems

Think about the answers to multiplication problems. In some situations, you can use general rules to get a sense if your answer is correct.

When you multiply a number (except for 0) by 1, the answer is *equal to* that number.

EXAMPLES $1 \times 23 = 23$ $3\frac{1}{2} \times 1 = 3\frac{1}{2}$ $1 \times \frac{2}{3} = \frac{2}{3}$

When you multiply a number greater than 0 by a proper fraction, the answer is always *less than* that number. In these problems you are finding *a part of* a number.

EXAMPLES $\frac{1}{2} \times 14 = 7$ $\frac{3}{4}$ of $12 = 9$ $\frac{2}{3} \times 1 = \frac{2}{3}$

**In each box write the symbol <, >, or = to make the statement true.**

1. $\frac{5}{6}$ of $18 \square 18$ $1 \times 4\frac{1}{2} \square 4\frac{1}{2}$ $\frac{3}{4} \times 2\frac{2}{5} \square 2\frac{2}{5}$

2. $9 \times 8\frac{2}{3} \square 8\frac{2}{3}$ $\frac{7}{12}$ of $\frac{3}{10} \square \frac{3}{10}$ $1 \times \frac{9}{16} \square \frac{9}{16}$

If the situations above do not apply, you can still get an estimation to check your answer. To estimate a multiplication problem with mixed numbers, round each mixed number to the nearest whole number. Then use the whole numbers to multiply.

EXAMPLE Estimate the answer to $5\frac{3}{4} \times 2$.

STEP 1 Round $5\frac{3}{4}$ to the nearest whole number. $5\frac{3}{4} \rightarrow 6$

STEP 2 Multiply the rounded numbers. $6 \times 2 = 12$

STEP 3 Find the exact answer. Notice how close the estimate, 12, is to the exact answer, $11\frac{1}{2}$. $5\frac{3}{4} \times 2 =$ $\frac{23}{\cancel{4}_2} \times \frac{\cancel{2}^1}{1} = \frac{23}{2} = \mathbf{11\frac{1}{2}}$

**First estimate each answer. Round each mixed number to the nearest whole number and find the product of the rounded numbers. Then find the exact answer.**

3. $6 \times 2\frac{3}{4} =$ $3\frac{1}{8} \times 4 =$ $5\frac{1}{3} \times 9 =$ $1\frac{5}{6} \times 8 =$

4. $4\frac{3}{5} \times 5 =$ $3 \times 6\frac{2}{9} =$ $12 \times 1\frac{7}{8} =$ $10\frac{1}{4} \times 6 =$

5. $2\frac{2}{9} \times 1\frac{7}{8} =$ $2\frac{2}{7} \times 5\frac{1}{4} =$ $4\frac{2}{3} \times 2\frac{7}{10} =$ $1\frac{1}{5} \times 4\frac{1}{6} =$

6. $2\frac{1}{4} \times 2\frac{2}{3} =$ $3\frac{3}{4} \times 4\frac{4}{9} =$ $5\frac{5}{6} \times 1\frac{13}{20} =$ $3\frac{3}{8} \times 2\frac{5}{9} =$

# Applying Your Multiplication Skills

Fraction multiplication problems sometimes ask you to find *a part of* something. In these problems, you multiply a fraction by another number. Because a part is always smaller than the whole, the answer is always *less than* the other number.

Other fraction multiplication problems may tell you the cost or the weight of *one* thing. Then the problem asks you to find the cost or the weight of *several* of those things. These answers are always *greater than* the unit cost or unit weight.

**For problems 1 to 8, solve each problem and write the correct label beside each answer. Then, to decide whether each answer is reasonable, write "part of" if you are finding a part of some amount. Or, if you are looking for more than one of some amount, round each mixed number to the nearest whole number. Then estimate the answer with the rounded numbers.**

1. Tim is a cabinet maker. He is replacing six warped shelves in a bookcase. Each shelf is $28\frac{1}{2}$ inches long. How many inches of shelving does he need for the project?

2. Janine's regular work week is 40 hours. One week she worked $\frac{4}{5}$ of that time. How many hours did she work that week?

3. The Leightons have to drive 276 miles to get from their house to their parents' house. If they have already driven $\frac{2}{3}$ of the distance, how far have they gone?

4. Adrienne takes home $2,980 every month. She spends $\frac{1}{4}$ of her take-home pay for rent on her 2-bedroom apartment. What is her monthly rent?

5. One cubic foot of water weighs $62\frac{1}{2}$ pounds. How much does $1\frac{1}{4}$ cubic feet of water weigh?

6. Verva drives 18 miles to her current job. A new job that she is considering is $1\frac{1}{2}$ times as far away as her current job. What is the distance to the new job?

7. Alex makes $270 per week working part-time in a gas station. He is able to save only $\frac{1}{10}$ of his weekly wages. How much does Alex save each week?

8. A board foot of cedar lumber costs $2. Find the cost of $5\frac{3}{4}$ board feet of the lumber.

**Read the next problems carefully. Some problems require operations other than multiplication.**

9. Robert works as an apprentice to an electrician. He makes $16 an hour. How much did he make on a job that took $3\frac{1}{4}$ hours to complete?

10. A tailor needs $3\frac{1}{6}$ yards of material to make a suit. How much material does he need to make three suits?

11. In the last problem, the tailor made the three suits from a piece of material $12\frac{1}{4}$ yards long. How much material was left?

In the weeks before Christmas, Serena worked $11\frac{1}{2}$ hours every day in a department store. Any hours that she works after 8 hours are considered overtime hours. Use this information to answer questions 12–15.

12. For the first 8 hours each day, Serena made $8.40 an hour. How much did she earn for those 8 hours?

13. For overtime work, Serena gets $1\frac{1}{2}$ times her regular wage. What does she earn for one hour of overtime work?

14. How much did Serena make each day for overtime work?

15. Find Serena's total income for an $11\frac{1}{2}$ hour day.

16. Marissa and Juan want to buy a condominium that costs $188,000. They plan to make a down payment of $\frac{1}{10}$ of the price of the condominium. How much is the down payment?

17. Juan's parents have offered to give them $9,000 toward the down payment. How much do Marissa and Juan have to take from their savings to complete the down payment?

Cassie is in charge of ticket sales at a community theater. On Saturday, 420 people attended a performance at the theater. Use this information to answer questions 18–20.

18. Of the people who attended, $\frac{2}{3}$ paid full price. How many people paid full price?

19. One-fourth of the people who attended the performance had student discounts or senior discounts. How many people had discounted tickets?

20. The rest of the people who attended the play had free, complimentary tickets. How many people had free tickets?

# Dividing Fractions by Fractions

The answer to a division problem is called the **quotient.** The number being divided is the **dividend,** and the number that divides into the dividend is the **divisor.**

$$\underset{\textbf{dividend}}{15} \div \underset{\textbf{divisor}}{3} = \underset{\textbf{quotient}}{5}$$

Notice that the whole number division problem, $15 \div 3 = 5$, has the same answer as the fraction multiplication problem, $15 \times \frac{1}{3} = 5$.

$$\frac{\overset{5}{\cancel{15}}}{1} \times \frac{1}{\underset{1}{\cancel{3}}} = \frac{5}{1} = 5$$

The fraction $\frac{1}{3}$ is called the **reciprocal** or the **inverse** of the improper fraction $\frac{3}{1}$. In division problems with fractions, you **invert** the divisor. To invert a fraction, write the denominator on top and the numerator below.

To divide a fraction, do the following steps:

**STEP 1** Invert the fraction to the right of the division sign (the divisor), and change the division sign (÷) to a multiplication sign (×).

**STEP 2** Follow the rules for multiplication.

EXAMPLE 1 **Suppose a gardener wants to divide her $\frac{1}{2}$-acre garden into $\frac{1}{8}$-acre sections. How many $\frac{1}{8}$-acre sections can she make?**

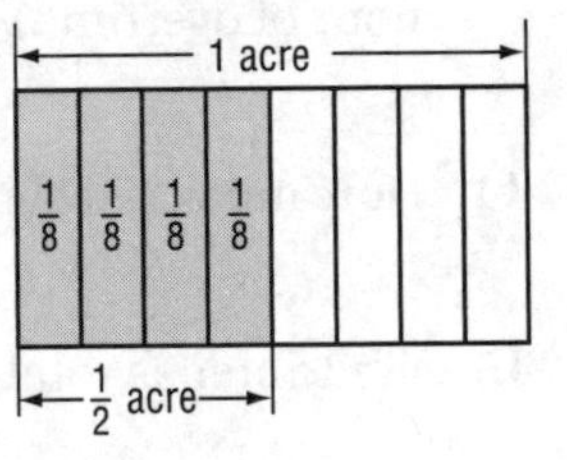


To find out how many sections she can make, divide $\frac{1}{2}$ by $\frac{1}{8}$.

**STEP 1** Invert the divisor $\frac{1}{8}$ to $\frac{8}{1}$ and change the ÷ sign to ×. $\quad \frac{1}{2} \div \frac{1}{8} =$

**STEP 2** Cancel and multiply across. $\quad \frac{1}{\underset{1}{\cancel{2}}} \times \frac{\overset{4}{\cancel{8}}}{1} = \frac{4}{1} = 4$

**STEP 3** Change the improper fraction to a whole number.
The gardener can make **four** $\frac{1}{8}$-acre sections.

ANSWER: **4 sections**

EXAMPLE 2 $\frac{3}{4} \div \frac{5}{8} =$

**STEP 1** Invert the divisor $\frac{5}{8}$ to $\frac{8}{5}$ and change the ÷ sign to ×. $\quad \frac{3}{4} \div \frac{5}{8} =$

**STEP 2** Cancel and multiply across. $\quad \frac{3}{\underset{1}{\cancel{4}}} \times \frac{\overset{2}{\cancel{8}}}{5} = \frac{6}{5} = \mathbf{1\frac{1}{5}}$

**STEP 3** Change the improper fraction to a mixed number.

ANSWER: $\mathbf{1\frac{1}{5}}$

**Divide and reduce.**

1. $\frac{3}{7} \div \frac{2}{5} =$ $\frac{4}{9} \div \frac{2}{3} =$ $\frac{5}{12} \div \frac{3}{4} =$ $\frac{7}{15} \div \frac{4}{5} =$

2. $\frac{11}{16} \div \frac{5}{8} =$ $\frac{8}{9} \div \frac{2}{9} =$ $\frac{7}{8} \div \frac{5}{6} =$ $\frac{21}{25} \div \frac{7}{10} =$

3. $\frac{14}{15} \div \frac{4}{9} =$ $\frac{35}{36} \div \frac{5}{12} =$ $\frac{9}{14} \div \frac{1}{2} =$ $\frac{6}{25} \div \frac{1}{5} =$

4. $\frac{1}{12} \div \frac{7}{9} =$ $\frac{4}{15} \div \frac{1}{15} =$ $\frac{49}{50} \div \frac{7}{10} =$ $\frac{8}{9} \div \frac{5}{12} =$

5. $\frac{5}{36} \div \frac{1}{4} =$ $\frac{13}{64} \div \frac{5}{8} =$ $\frac{17}{48} \div \frac{1}{24} =$ $\frac{7}{16} \div \frac{3}{8} =$

6. $\frac{3}{10} \div \frac{6}{7} =$ $\frac{5}{12} \div \frac{3}{16} =$ $\frac{2}{3} \div \frac{25}{36} =$ $\frac{9}{20} \div \frac{3}{4} =$

# Dividing Whole Numbers by Fractions

The owner of a 3-acre parcel of land wants to divide the land into $\frac{3}{4}$-acre lots for resale. How many $\frac{3}{4}$-acre lots can be made from the large parcel? To answer this question, you find out how many $\frac{3}{4}$'s there are in 3. In other words, divide 3 by $\frac{3}{4}$.

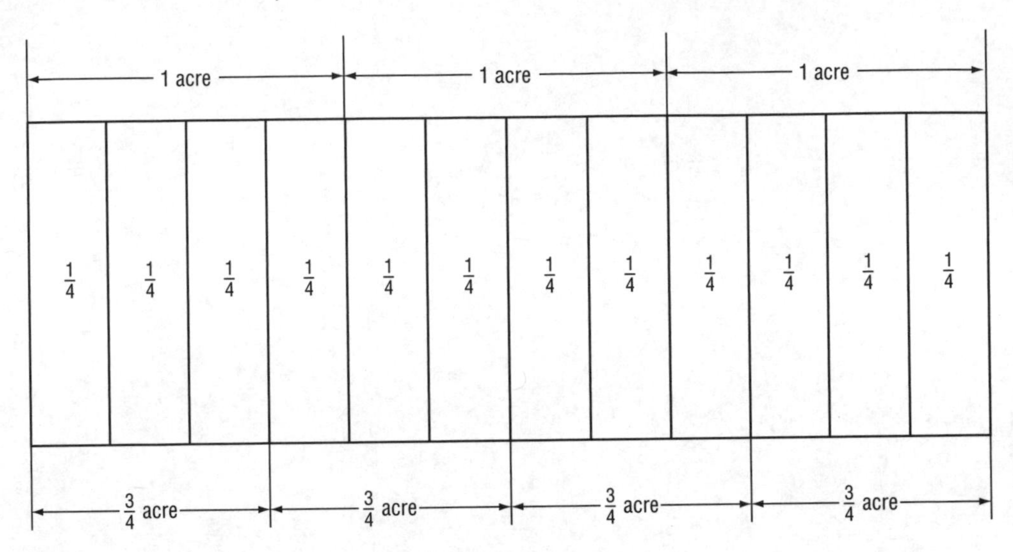


The illustration shows that the owner can make four $\frac{3}{4}$-acre lots.

When dividing a whole number by a fraction, write the whole number as an improper fraction with a denominator of 1.

**EXAMPLE 1** $3 \div \frac{3}{4} =$

**STEP 1** Write 3 as the improper fraction $\frac{3}{1}$.

**STEP 2** Invert $\frac{3}{4}$ to $\frac{4}{3}$ and change the ÷ sign to ×.

**STEP 3** Cancel and multiply across.

**STEP 4** Change the improper fraction to a whole number.

$$\frac{3}{1} \div \frac{3}{4} =$$

$$\frac{\cancel{3}^{1}}{1} \times \frac{4}{\cancel{3}_{1}} = \frac{4}{1} = \mathbf{4}$$

Notice the answer in the last example. You divided the whole number 3 by a fraction that is smaller than 1. That fraction, $\frac{3}{4}$, divides into 3 *more than* three times.

**EXAMPLE 2** $8 \div \frac{6}{7} =$

**STEP 1** Write 8 as the improper fraction $\frac{8}{1}$.

**STEP 2** Invert $\frac{6}{7}$ to $\frac{7}{6}$ and change the ÷ sign to ×.

**STEP 3** Cancel and multiply across.

**STEP 4** Change the improper fraction to a mixed number.

$$\frac{8}{1} \div \frac{6}{7} =$$

$$\frac{\cancel{8}^{4}}{1} \times \frac{7}{\cancel{6}_{3}} = \frac{28}{3} = \mathbf{9\frac{1}{3}}$$

**Divide and reduce.**

**1.** $12 \div \frac{2}{5} =$ $9 \div \frac{1}{3} =$ $5 \div \frac{3}{4} =$ $6 \div \frac{2}{3} =$

**2.** $8 \div \frac{3}{8} =$ $25 \div \frac{5}{6} =$ $17 \div \frac{1}{2} =$ $32 \div \frac{8}{9} =$

**3.** $45 \div \frac{9}{10} =$ $49 \div \frac{7}{12} =$ $36 \div \frac{24}{25} =$ $54 \div \frac{9}{20} =$

**4.** $48 \div \frac{8}{15} =$ $27 \div \frac{3}{4} =$ $16 \div \frac{3}{5} =$ $18 \div \frac{9}{16} =$

**5.** $16 \div \frac{4}{15} =$ $45 \div \frac{3}{20} =$ $15 \div \frac{5}{16} =$ $12 \div \frac{9}{10} =$

**6.** $20 \div \frac{5}{8} =$ $5 \div \frac{15}{16} =$ $7 \div \frac{5}{12} =$ $16 \div \frac{24}{25} =$

# Dividing Fractions by Whole Numbers

Think about dividing one more piece of land. If a $\frac{3}{4}$-acre parcel is divided into two equal pieces, what is the size of each parcel? To answer this question, you divide $\frac{3}{4}$ by 2.

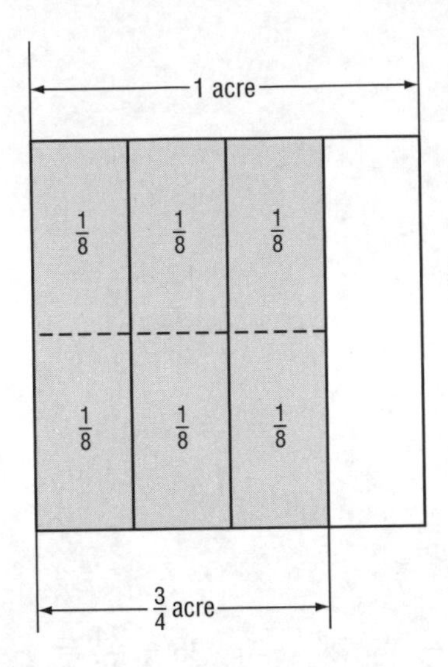


When dividing a fraction by a whole number, first write the whole number as an improper fraction with a denominator of 1. Then invert that fraction and multiply.

EXAMPLE 1 $\frac{3}{4} \div 2 =$

**STEP 1** Write 2 as the improper fraction $\frac{2}{1}$. $\frac{3}{4} \div \frac{2}{1} =$

**STEP 2** Invert $\frac{2}{1}$ to $\frac{1}{2}$ and change the $\div$ sign to $\times$. $\frac{3}{4} \times \frac{1}{2} = \mathbf{\frac{3}{8}}$

**STEP 3** Multiply across.

**ANSWER:** Each smaller parcel is $\mathbf{\frac{3}{8}}$**-acre.**

In the last example, notice that dividing by 2 is the same as multiplying by $\frac{1}{2}$. In other words, dividing a number by 2 is the same as finding *one-half of the number.*

EXAMPLE 2 $\frac{8}{9} \div 6 =$

**STEP 1** Write 6 as the improper fraction $\frac{6}{1}$. $\frac{8}{9} \div \frac{6}{1} =$

**STEP 2** Invert $\frac{6}{1}$ to $\frac{1}{6}$ and change the $\div$ sign to $\times$. $\frac{\not{8}^{4}}{9} \times \frac{1}{\not{6}_{3}} = \mathbf{\frac{4}{27}}$

**STEP 3** Cancel and multiply across.

**Divide and reduce.**

| | | | |
|---|---|---|---|
| **1.** $\frac{4}{5} \div 4 =$ | $\frac{2}{3} \div 6 =$ | $\frac{3}{16} \div 9 =$ | $\frac{1}{4} \div 2 =$ |
| **2.** $\frac{1}{3} \div 12 =$ | $\frac{3}{4} \div 7 =$ | $\frac{5}{6} \div 10 =$ | $\frac{9}{10} \div 3 =$ |
| **3.** $\frac{1}{2} \div 11 =$ | $\frac{3}{5} \div 15 =$ | $\frac{5}{8} \div 5 =$ | $\frac{12}{25} \div 18 =$ |
| **4.** $\frac{3}{4} \div 24 =$ | $\frac{9}{10} \div 36 =$ | $\frac{14}{15} \div 35 =$ | $\frac{24}{25} \div 40 =$ |
| **5.** $\frac{3}{5} \div 9 =$ | $\frac{9}{10} \div 6 =$ | $\frac{15}{16} \div 12 =$ | $\frac{7}{8} \div 28 =$ |
| **6.** $\frac{4}{9} \div 44 =$ | $\frac{15}{16} \div 20 =$ | $\frac{5}{18} \div 15 =$ | $\frac{9}{20} \div 12 =$ |

# Dividing with Mixed Numbers

To divide with mixed numbers, first change every mixed number to an improper fraction (see page 21). Also, be sure to write whole numbers as fractions over 1. Then invert the fraction to the right of the division sign and finish the problems as on pages 48, 50, and 52.

EXAMPLE $2\frac{1}{3} \div \frac{1}{4} =$

**STEP 1** Write $2\frac{1}{3}$ as the improper fraction $\frac{7}{3}$. $\frac{7}{3} \div \frac{1}{4} =$

**STEP 2** Invert $\frac{1}{4}$ to $\frac{4}{1}$ and change the ÷ sign to ×. $\frac{7}{3} \times \frac{4}{1} = \frac{28}{3} = \mathbf{9\frac{1}{3}}$

**STEP 3** Since nothing can be canceled, multiply across.

**STEP 4** Change the improper fraction to a mixed number.

**Divide and reduce.**

**1.** $1\frac{1}{2} \div \frac{3}{4} =$ $1\frac{2}{3} \div \frac{2}{3} =$ $2\frac{3}{4} \div \frac{5}{8} =$ $4\frac{1}{3} \div \frac{2}{9} =$

**2.** $2\frac{2}{5} \div 6 =$ $3\frac{1}{3} \div 4 =$ $1\frac{5}{7} \div 9 =$ $2\frac{2}{9} \div 15 =$

**3.** $\frac{5}{8} \div 1\frac{1}{4} =$ $\frac{14}{15} \div 1\frac{1}{6} =$ $\frac{7}{12} \div 2\frac{1}{2} =$ $\frac{9}{16} \div 3\frac{3}{4} =$

**4.** $12 \div 1\frac{3}{5} =$ $20 \div 2\frac{2}{7} =$ $9 \div 1\frac{7}{8} =$ $14 \div 2\frac{3}{16} =$

**5.** $3\frac{3}{4} \div 1\frac{1}{8} =$ $4\frac{1}{2} \div 1\frac{1}{6} =$ $2\frac{3}{4} \div 1\frac{7}{8} =$ $1\frac{7}{9} \div 2\frac{2}{9} =$

**6.** $5\frac{1}{4} \div 4\frac{2}{3} =$ $6\frac{1}{2} \div 3\frac{1}{4} =$ $5\frac{3}{5} \div 3\frac{1}{2} =$ $4\frac{3}{8} \div 1\frac{9}{16} =$

**7.** $5\frac{2}{3} \div 1\frac{8}{9} =$ $4\frac{2}{5} \div 8\frac{4}{5} =$ $3\frac{5}{9} \div 1\frac{13}{15} =$ $10\frac{2}{3} \div 2\frac{2}{3} =$

**8.** $3\frac{3}{5} \div 1\frac{7}{20} =$ $6\frac{3}{5} \div 2\frac{1}{5} =$ $1\frac{3}{8} \div 1\frac{1}{6} =$ $1\frac{1}{4} \div 3\frac{3}{4} =$

**9.** $4\frac{2}{3} \div 1\frac{3}{5} =$ $6\frac{7}{8} \div 5\frac{1}{4} =$ $9\frac{7}{10} \div 1\frac{4}{5} =$ $8\frac{2}{3} \div 1\frac{4}{9} =$

**10.** $3\frac{3}{5} \div 1\frac{1}{8} =$ $3\frac{5}{9} \div 1\frac{1}{3} =$ $5\frac{5}{6} \div 2\frac{1}{12} =$ $10\frac{5}{8} \div 4\frac{1}{2} =$

# Finding a Number When a Fraction of It Is Given

There is a kind of division problem that is sometimes hard to recognize. Think about the question: $\frac{1}{2}$ of what number is 12? Without using pencil and paper, you can probably come up with the answer 24. You know that $\frac{1}{2}$ of 24 is 12.

To solve the problem, you find a solution to the statement $\frac{1}{2} \times ? = 12$. The statement asks you to find the missing number in a multiplication problem. Division is the opposite operation of multiplication. To find the missing number, divide 12 by $\frac{1}{2}$.

You will learn more about opposite operations when you study algebra.

**EXAMPLE** $\frac{1}{2}$ of what number is 12?

**STEP 1** Write 12 as the improper fraction $\frac{12}{1}$.

**STEP 2** Invert $\frac{1}{2}$ to $\frac{2}{1}$ and change the ÷ sign to ×.

**STEP 3** Multiply across.

**STEP 4** Change the improper fraction to a whole number.

$\frac{12}{1} \div \frac{1}{2} =$

$\frac{12}{1} \times \frac{2}{1} = \frac{24}{1} = \mathbf{24}$

**Solve.**

**1.** $\frac{1}{3}$ of what number is 5? $\frac{1}{4}$ of what number is 22?

**2.** $\frac{2}{3}$ of what number is 18? $\frac{5}{6}$ of what number is 60?

**3.** $\frac{3}{4}$ of what number is 15? $\frac{5}{8}$ of what number is 40?

**4.** $\frac{2}{5}$ of what number is 30? $\frac{7}{12}$ of what number is 14?

**5.** $\frac{3}{8}$ of what number is 36? $\frac{1}{6}$ of what number is 11?

**6.** $\frac{4}{5}$ of what number is 80? $\frac{8}{9}$ of what number is 32?

**7.** $\frac{3}{10}$ of what number is 45? $\frac{7}{12}$ of what number is 35?

# Estimating Division Answers

Remember that in the problem $6 \div \frac{1}{2} = 12$, the number 6 is the dividend, $\frac{1}{2}$ is the divisor, and 12 is the quotient. The problem states that "6 divided by $\frac{1}{2}$ is 12." In some situations, you can use general rules to get a sense if your answer is correct.

**When a number greater than 0 is divided by a proper fraction, the quotient is always *greater than* the dividend.**

> When a whole number or a mixed number is divided by a proper fraction, the answer is always *greater than* the whole number or the mixed number.

EXAMPLES $6 \div \frac{1}{2} = 12$ $\quad 2\frac{1}{2} \div \frac{1}{4} = 10$ $\quad 15 \div \frac{3}{4} = 20$

**When a number greater than 0 is divided by a larger number, the quotient is always *a fraction*.**

> When a proper fraction is divided by a whole number or a mixed number, the answer is always a *smaller* fraction

EXAMPLES $\frac{1}{2} \div 6 = \frac{1}{12}$ $\quad \frac{3}{4} \div 1\frac{1}{2} = \frac{1}{2}$ $\quad \frac{2}{3} \div 5 = \frac{2}{15}$

**In each box, write the symbol < (less than) or > (greater than) to make the statement true.**

1. $\frac{5}{6} \div 2 \square \frac{5}{6}$ $\quad 1\frac{3}{8} \div \frac{1}{2} \square 1\frac{3}{8}$ $\quad 12 \div \frac{9}{10} \square 12$
2. $9\frac{2}{3} \div \frac{3}{4} \square 9\frac{2}{3}$ $\quad \frac{11}{12} \div 4\frac{1}{6} \square \frac{11}{12}$ $\quad 1 \div \frac{9}{16} \square 1$

If the situations above do not apply, you can still get an estimate. To estimate fraction division problems with mixed numbers, round any mixed number to the nearest whole number. Then divide the whole numbers to estimate the answer.

EXAMPLE Estimate an answer to the problem $12 \div 2\frac{2}{3} =$

**STEP 1** Round $2\frac{2}{3}$ to the nearest whole number. $2\frac{2}{3} \rightarrow 3$

**STEP 2** Divide the rounded numbers. $12 \div 3 = \mathbf{4}$

The exact answer is $12 \div 2\frac{2}{3} = \frac{12}{1} \div \frac{8}{3} = \frac{\overset{3}{\cancel{12}}}{1} \times \frac{3}{\underset{2}{\cancel{8}}} = \frac{9}{2} = 4\frac{1}{2}$.

**Solve each problem. Then estimate the answer by rounding any mixed number to the nearest whole number and dividing the rounded numbers.**

3. $1\frac{3}{4} \div \frac{1}{2} =$ $\quad 4\frac{4}{5} \div \frac{2}{5} =$ $\quad 2\frac{2}{3} \div 4 =$
4. $5 \div 1\frac{7}{8} =$ $\quad 11 \div 4\frac{2}{5} =$ $\quad 7 \div 3\frac{1}{9} =$
5. $3\frac{3}{4} \div 2\frac{1}{2} =$ $\quad 2\frac{2}{9} \div 5\frac{1}{3} =$ $\quad 6\frac{7}{8} \div 4\frac{1}{6} =$
6. $5\frac{1}{3} \div 10 =$ $\quad 3\frac{1}{2} \div \frac{7}{10} =$ $\quad 2\frac{4}{5} \div 4 =$

# Applying Your Division Skills

Word problems with division of fractions are tricky because you need to invert the correct number. Remember that the amount being *divided, cut,* or *shared* must be written to the *left* of the ÷ sign. For example, if $5\frac{1}{2}$ pounds of peanuts are to be divided equally among three people, write $5\frac{1}{2}$ first. The problem becomes $5\frac{1}{2} \div 3$. The number on the *right* gets inverted.

**For problems 1 to 8, solve and write the correct label, such as pounds or inches, next to each answer. Reduce each answer to lowest terms. Then round fractions and mixed numbers to the nearest whole number and use the rounded numbers to estimate each answer.**

**1.** How many pieces of wood each $7\frac{1}{2}$ inches long can be cut from a board that is 45 inches long?

**2.** Sandy baked $4\frac{1}{2}$ pounds of cookies. She divided the cookies equally among two friends and herself. How many pounds of cookies did each person get?

**3.** Fabrizio is a tailor. He needs $3\frac{2}{3}$ yards of material to make a suit. How many suits can he make from 22 yards of material?

**4.** How many $1\frac{1}{2}$-pound loaves of bread can be made from 9 pounds of dough?

**5.** How many $\frac{3}{4}$-pound cans of tomatoes can be filled with 24 pounds of tomatoes?

**6.** To make picture frames, Carlos plans to cut a piece of wooden molding 75 inches long into small strips each $8\frac{1}{3}$ inches long. Assuming there is no waste, how many strips can he cut from the long piece?

**7.** Sarah needs $2\frac{1}{4}$ yards of material to make a dress for her daughter. How many dresses can she make from $10\frac{1}{2}$ yards of material?

**8.** Adriano wants to share $5\frac{1}{2}$ pounds of peanuts equally among three people. Find the weight of peanuts each person will get.

**Read each problem carefully. Some problems require operations other than division.**

9. Rashid repairs computers for a large bank. On average he can make a repair in $\frac{3}{4}$ of an hour. How many repairs can he make in a $7\frac{1}{2}$-hour workday?

10. On average, how many repairs can Rashid, in the last problem, complete in a five-day work week?

11. On a snowy evening, only $\frac{2}{3}$ of the registered students in Bob's chess class came to school. Eighteen students attended that night. How many students are registered in Bob's class?

12. In the last problem, how many of the registered students were absent on the snowy evening?

13. A safety inspector asked workers in a factory if they thought conditions were adequate. Ninety of the workers answered that they thought conditions were satisfactory. They represent $\frac{3}{5}$ of all the workers who were questioned. How many workers were questioned?

14. In the last problem, how many of the workers did not think that safety conditions were adequate?

15. On the first day of the fishing season, Jed caught $10\frac{1}{2}$ pounds of trout. He decided to keep half of the amount for himself. How many pounds of trout did he keep?

16. In the last problem, Jed gave the remaining trout to three friends. If they each got the same weight, how much trout did each receive?

17. The budget of $\$1\frac{1}{2}$ million to renovate a school gym will be shared equally between the state and the local community. How much does the local community have to pay for the renovation?

18. In the last problem, the school expects to get $\frac{1}{3}$ of the amount that the community owes for the gym renovation from donations. The rest will come from taxes. How much will the community have to raise in taxes?

# Fractions Review

This review covers the material you have studied so far in this book. When you finish, check your answers at the back of the book.

**1.** A yard contains 36 inches. 21 inches is what fraction of a yard?

**2.** Which fractions in this list are less than $\frac{1}{2}$? $\frac{3}{6}, \frac{7}{15}, \frac{4}{9}, \frac{5}{8}$

**3.** Reduce $\frac{18}{32}$ to lowest terms.

**4.** Write $\frac{50}{12}$ as a mixed number.

**5.** Write $11\frac{5}{8}$ as an improper fraction.

**6.** Change $\frac{3}{4}$ to twelfths.

**7.** $9\frac{7}{8} + 6\frac{5}{8} + 3\frac{3}{8} =$

**8.** $4\frac{3}{5} + 8\frac{1}{2} + 7\frac{2}{3} =$

**9.** Without finding the exact sum, tell whether $\frac{5}{8} + \frac{5}{9}$ is less than 1, greater than 1, or equal to 1.

**10.** Mr. Gutierrez usually takes $\frac{3}{4}$ hour to drive home from work. Because of a traffic jam, he took an extra $1\frac{2}{3}$ hours to get home one night. How long did that ride take him?

**11.** Find the combined weight of three packages that weigh $5\frac{1}{2}$ pounds, $4\frac{7}{16}$ pounds, and $3\frac{3}{8}$ pounds.

**12.** $10\frac{7}{8} - 5\frac{2}{5} =$

**13.** $15\frac{2}{9} - 9\frac{7}{9} =$

**14.** $13\frac{2}{9} - 4\frac{5}{6} =$

**15.** From a 2-pound box of chocolates, Rachel ate $1\frac{1}{4}$ pounds. What was the weight of the remaining chocolates?

**16.** The distance from Ellen's home to her school is $4\frac{1}{3}$ miles. If she has already traveled $2\frac{1}{2}$ miles, how far does she have to go?

17. $\frac{2}{3} \times \frac{4}{5} =$

18. $\frac{3}{8} \times \frac{14}{15} \times \frac{1}{2} =$

19. $2\frac{1}{4} \times 1\frac{5}{9} =$

20. In the problem $2\frac{4}{5} \times 6\frac{1}{4}$, round each number to the nearest whole number. Then multiply.

21. What is the total weight of four cartons if each carton weighs $16\frac{1}{4}$ pounds?

22. Olivia walks $3\frac{1}{2}$ miles per hour. How far can she walk in $2\frac{1}{4}$ hours?

23. $\frac{8}{15} \div \frac{7}{12} =$

24. $14 \div \frac{4}{9} =$

25. $\frac{9}{10} \div 6 =$

26. $3\frac{1}{8} \div 5\frac{5}{6} =$

27. $\frac{4}{5}$ of what number is 60?

28. A carpenter needs $3\frac{1}{2}$ yards of lumber to build a bookcase. How many bookcases can he build from 21 yards of lumber?

29. For a $7\frac{1}{2}$-hour day of work, Ed makes $120. How much does he make in 1 hour?

## FRACTIONS REVIEW CHART

If you missed more than one problem on any group below, review the practice pages for those problems. Then redo the problems you got wrong before going on to the Decimal Skills Inventory. If you had a passing score, redo any problem you missed and begin the Decimal Skills Inventory on page 62.

| Problem Numbers | Skill Area | Practice Pages |
|---|---|---|
| 1, 2, 3, 4, 5, 6 | understanding fractions | 11–21 |
| 7, 8, 9, 10, 11 | adding fractions | 22–30 |
| 12, 13, 14, 15, 16 | subtracting fractions | 31–39 |
| 17, 18, 19, 20, 21, 22 | multiplying fractions | 40–47 |
| 23, 24, 25, 26, 27, 28, 29 | dividing fractions | 48–59 |

# DECIMALS

## Decimal Skills Inventory

This inventory will tell you whether you need to work through the decimals section of this book. Do all the problems that you can. Work carefully and check your answers, but do not use outside help. Correct answers are listed by page number at the back of the book.

**1.** Write seventy-six thousandths as a decimal.

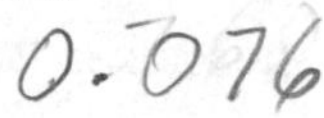

**2.** Write 3.04 as a mixed number and reduce.

**3.** Write $\frac{5}{12}$ as a decimal.

**4.** Which is larger, 0.076 or 0.08?

**5.** Rewrite the following list in order from smallest to largest:

0.087, 0.7, 0.08, 0.07

**6.** Round 2.093 to the nearest tenth.

**7.** $4.3 + 0.079 + 0.06 =$

**8.** $21 + 2.86 + 0.093 =$

**9.** Find the total weight of three packages that weigh 2.3 pounds, 4.15 pounds, and 5.65 pounds.

**10.** At 5:00 on a June morning, the temperature was 62.4°. By 2:00 in the afternoon, the temperature had risen 21.7°. What was the temperature at 2:00 that afternoon?

**11.** $9.6 - 0.457 =$

**12.** $12 - 0.37 =$

**13.** From a piece of yarn 3 yards long, Sylvia cut a piece 1.85 yards long. How long was the remaining piece?

**14.** When Fred began driving Saturday morning, his mileage gauge read 8,249.6 miles. When he returned home that night, it read 8,536.4 miles. How far did Fred drive that day?

**15.** $2.07 \times 5.3 =$

**16.** $0.038 \times 0.6 =$

**17.** $4.92 \times 10 =$

**18.** Diane drove her car at an average speed of 45 miles per hour for 3.2 hours. How far did she drive in that time?

**19.** At $12.60 an hour, how much does a worker make in 7.25 hours?

**20.** $12.04 \div 14 =$

**21.** $23.36 \div 3.2 =$

**22.** $5.55 \div 0.015 =$

**23.** $15 \div 0.075 =$

**24.** $49.6 \div 100 =$

**25.** If 1.4 pounds of beef cost $3.64, how much does 1 pound of beef cost?

**26.** One yard contains 0.914 meter. How many yards are there in 4.57 meters?

## DECIMALS INVENTORY CHART

If you missed more than one problem on any group below, work through the practice pages for that group. Then redo the problems you got wrong on the Decimals Inventory Test. If you had a passing score on all five groups of problems, redo any problem you missed and begin the Percent Skills Inventory on page 99.

| Problem Numbers | Skill Area | Practice Pages |
|---|---|---|
| 1, 2, 3, 4, 5, 6 | understanding decimals | 64–73 |
| 7, 8, 9, 10 | adding decimals | 74–76 |
| 11, 12, 13, 14 | subtracting decimals | 77–80 |
| 15, 16, 17, 18, 19 | multiplying decimals | 81–87 |
| 20, 21, 22, 23, 24, 25, 26 | dividing decimals | 88–96 |

# Understanding Decimals

A decimal is a type of fraction that you work with every day. The following numbers have decimal fractions: $3.75, $64.08, and $1.20. Each number represents dollars and cents. The point in each number separates whole dollar amounts from fractions of a dollar.

There are 100 cents in a dollar. One cent is *one hundredth* of a dollar. The number $3.75 represents three whole dollars and $\frac{75}{100}$ of a dollar. Digits to the left of the decimal point are in whole number places. Digits to the right are in decimal places.

Our money system has just two decimal places. They are dimes or tenths of a dollar and pennies or hundredths of a dollar. Study the chart below. It tells the names of seven whole number places and six decimal places.

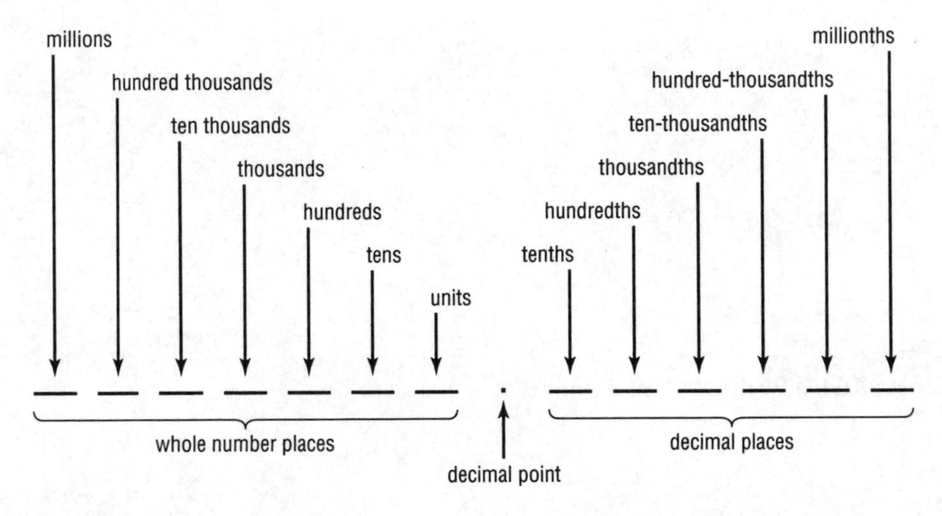


Notice how similar the names are on each side of the decimal point. The first place to the left of the units (or ones) is *tens.* The first place to the right of the units is *tenths.* The third place to the left of units is *thousands.* The third place to the right of units is *thousandths.*

The *–ths* at the end of each decimal place name means that the value of a digit in these places is a fraction of one unit or one whole.

**EXAMPLE 1** **In the decimal 0.135, which digit is in the *thousandths* place?**

Thousandths are the third place to the right of the decimal point. In 0.135, the digit 5 is in the thousandths place.

**EXAMPLE 2** **What is the value of the digit 3 in the decimal 0.135?**

Since 3 is in the second decimal place, it has a value of 3 hundredths.

A number with digits on both sides of the decimal point is sometimes called a **mixed decimal.**

**EXAMPLE 3** **In the number 19.64, which digits are in the decimal places?**

The digits 6 and 4 are in the decimal places.

**For each number, underline the digit that is in the place named.**

1. the *tenths* place

   1.3 0.5864 297.18 43.682

2. the *dimes* (or *tenths*) place

   $0.43 $107.21 $6.39 $80.10

3. the *hundredths* place

   0.56 2.974 0.1389 63.4528

4. the *pennies* (or *hundredths*) place

   $3.59 $10.67 $1.026 $295.43

5. the *thousandths* place

   1.234 0.18956 0.04107 0.00385

6. the *ten-thousandths* place

   15.46935 0.02684 3.14159 0.00067

**Circle the correct answer for each question.**

7. Which of the following tells the value of the digit 9 in the number 2.936?

   9 tenths 9 hundredths 9 thousandths 9 ten-thousandths

8. Which of the following tells the value of the digit 7 in the number 12.047?

   7 tenths 7 hundredths 7 thousandths 7 ten-thousandths

# Reading Decimals

To read a decimal, first read it as though it were a whole number. Then give it a decimal name according to the number of decimal places. Remember that the number of places to the right of the decimal point tells you the decimal name. Use the chart on page 64 as a reminder. In the following examples, notice how zeros hold places. With mixed decimals, the word *and* separates whole numbers from decimal fractions.

EXAMPLES

| Decimal | Number of Decimal Places | Words |
|---|---|---|
| 0.06 | two | *six hundredths* |
| 0.0042 | four | *forty-two ten-thousandths* |
| 7.003 | three | *seven and three thousandths* |
| 19.5 | one | *nineteen and five tenths* |

**For problems 1 through 6, fill in the blanks with the correct decimal name.**

**1.** 0.8 = eight ________ 0.3 = three ________ 0.06 = six ________

**2.** 0.17 = seventeen ________ 0.09 = nine ________ 0.005 = five ________

**3.** 0.023 = twenty-three ________ 0.0028 = twenty-eight ________

**4.** 0.308 = three hundred eight ________ 0.0006 = six ________

**5.** 4.02 = four and two ________ 20.012 = twenty and twelve ________

**6.** 90.3 = ninety and three ________ 5.0041 = five and forty-one ________

**For problems 7 through 11, write each decimal or mixed decimal in words.**

**7.** 0.5 = 2.9 =

**8.** 0.07 = 3.12 =

**9.** 0.016 = 1.003 =

**10.** 0.009 = 7.021 =

**11.** 0.0016 = 10.402 =

# Writing Decimals

To write decimals from words, be sure that you have the correct number of decimal places. Use zeros to hold places where necessary. Remember, again, that the word *and* separates whole numbers from decimal fractions in mixed decimals. Study the following examples carefully. Watch where zeros hold places.

EXAMPLES

| Words | Number of Decimal Places | Decimal |
|---|---|---|
| *eight hundredths* | two | 0.08 |
| *thirty-six thousandths* | three | 0.036 |
| *fifty-one millionths* | six | 0.000051 |
| *two and five thousandths* | three | 2.005 |
| *twelve and nine tenths* | one | 12.9 |

**Write the following as a decimal or a mixed decimal.**

1. seven tenths — four and nine tenths
2. sixty-three hundredths — twenty-two thousandths
3. two hundredths — eighty-five hundredths
4. four tenths — nineteen thousandths
5. twelve and three tenths — forty-one hundredths
6. one and five hundredths — two hundred six thousandths
7. forty-eight and nine tenths — eight thousandths
8. three hundred four thousandths — eleven and seven hundredths
9. fifteen ten-thousandths — four and thirteen thousandths
10. two hundred thirty and one tenth — seventy-one ten-thousandths
11. forty-seven thousandths — eight and two ten-thousandths
12. five hundred thousandths — nineteen millionths

# Getting Rid of Unnecessary Zeros

Think about the zeros in the number 020.060 and decide whether each zero is necessary. The number is *twenty and six hundredths.*

The zero *left* of the digit 2 is unnecessary because without it, 2 remains in the tens place.
The zero *right* of 2 is necessary because it keeps the 2 in the tens place.
The zero *left* of 6 is necessary because it keeps 6 in the hundredths place.
The zero *right* of 6 is unnecessary because without it, 6 remains in the hundredths place.

The number can be correctly rewritten as 20.06.

**Note:** The sentence above ends with a period. Do not confuse the period at the end of a sentence with a decimal point.

A decimal with no whole number is often written with a zero in the units place. The decimals .8 and 0.8 are both correct forms for *eight tenths.*

**For each number, choose the correctly rewritten number.**

**1.** 5.0060

a. 50.6 **b.** 50.06 **c.** 5.006 **d.** 5.06

**2.** 003.1050

**a.** 3.105 **b.** 30.105 **c.** 3.15 **d.** 30.015

**3.** 0700.40

**a.** 70.04 **b.** 70.4 **c.** 700.04 **d.** 700.4

**4.** 0040.0920

**a.** 40.92 **b.** 40.092 **c.** 4.092 **d.** 4.92

**5.** 00.1060

**a.** .106 **b.** 0.106 **c.** 0.16 **d.** both a and b

**6.** 0.03400

**a.** 0.34 **b.** 0.034 **c.** .34 **d.** both b and c

# Changing Decimals to Fractions

To change a decimal to a fraction (or to change a mixed decimal to a mixed number), write the digits in the decimal as the numerator. Write the denominator according to the number of decimal places. Then reduce the fraction.

**EXAMPLE 1** Write 0.24 as a common fraction.

**STEP 1** Write 24 as the numerator. Since two places means hundredths, write 100 as the denominator.

**STEP 2** Reduce by dividing 24 and 100 by 4.

$$\frac{24 \div 4}{100 \div 4} = \mathbf{\frac{6}{25}}$$

**EXAMPLE 2** Write 9.015 as a mixed number.

**STEP 1** Write 9 as the whole number and 15 as the numerator. Since three places means thousandths, write 1,000 as the denominator.

**STEP 2** Reduce by dividing 15 and 1,000 by 5.

$$9\frac{15 \div 5}{1000 \div 5} = \mathbf{9\frac{3}{200}}$$

---

**Write each number as a common fraction or a mixed number and reduce.**

| | | |
|---|---|---|
| 1. 0.08 = | 0.375 = | 0.0048 = |
| **2.** 3.6 = | 9.86 = | 10.002 = |
| **3.** 0.085 = | 5.08 = | 0.0025 = |
| **4.** 7.2 = | 0.15 = | 8.16 = |
| **5.** 0.00324 = | 19.0786 = | 123.462 = |
| **6.** 16.00004 = | 7.22 = | 3.000008 = |
| **7.** 2,036.8 = | 48.02 = | 3.075 = |

# Changing Fractions to Decimals

Remember that a fraction can be understood as a division problem. To change a fraction to a decimal, divide the denominator into the numerator. To divide, add a decimal point and zeros to the numerator. Usually two zeros are enough. Then bring the point up in the answer.

**EXAMPLE 1** Write $\frac{1}{2}$ as a decimal.

**STEP 1** Divide 2 into 1. Add a decimal point and a zero to 1.

$$2\overline{)1.0}\quad \mathbf{0.5}$$

**STEP 2** Divide and bring up the decimal point.

**EXAMPLE 2** Write $\frac{3}{20}$ as a decimal.

**STEP 1** Divide 20 into 3. Add a decimal point and two zeros to 3.

$$20\overline{)3.00}\quad \mathbf{0.15}$$

**STEP 2** Divide and bring up the decimal point.

**EXAMPLE 3** Write $1\frac{2}{3}$ as a mixed decimal.

**STEP 1** Change $1\frac{2}{3}$ to an improper fraction. $1\frac{2}{3} = \frac{5}{3}$

$$3\overline{)5.00}\quad \mathbf{1.66\tfrac{2}{3}}$$

**STEP 2** Divide 3 into 5. Add a decimal point and two zeros to 5.

**STEP 3** Divide and bring up the decimal point.

In the last example, the division will not come out evenly no matter how many zeros you add. After two places, write the remainder as a fraction over the number you divided by.

**Write each fraction or mixed number as a decimal or a mixed decimal.**

**1.** $\frac{1}{4} =$ $\frac{2}{5} =$ $\frac{5}{8} =$ $\frac{7}{2} =$

**2.** $\frac{2}{9} =$ $\frac{6}{25} =$ $\frac{1}{6} =$ $\frac{11}{8} =$

**3.** $\frac{5}{6} =$ $\frac{3}{10} =$ $\frac{4}{7} =$ $\frac{11}{6} =$

**4.** $\frac{3}{5} =$ $\frac{9}{10} =$ $\frac{1}{12} =$ $\frac{9}{4} =$

**5.** $\frac{2}{15} =$ $\frac{3}{4} =$ $\frac{1}{8} =$ $\frac{13}{10} =$

# Comparing Decimals

When you look at a group of decimal fractions, it is sometimes difficult to tell which decimal is the largest. To compare decimals, give each decimal the same number of places by adding zeros. This is the same as giving each decimal fraction a common denominator. The zeros you add do not change the value of the decimals.

EXAMPLE 1 **Which is larger, 0.07 or 0.2?**

STEP 1 Add a zero to 0.2 so that both decimals have two places. — 0.07 or 0.20

STEP 2 Since twenty hundredths is larger than seven hundredths, 0.2 is larger. — **0.2 is larger**

EXAMPLE 2 **Arrange the following list of decimals in order from smallest to largest: 0.8, 0.08, 0.088, 0.808**

STEP 1 Add zeros so that each decimal has the same number of places. — 0.800, 0.080, 0.088, 0.808

STEP 2 Compare and arrange the decimals in order from smallest to largest. — **0.08, 0.088, 0.8, 0.808**

Notice that the extra zeros are not written in the final list.

**In each pair, tell which decimal is larger.**

**1.** 0.04 or 0.008    0.9 or 0.99    0.67 or 0.707

**2.** 0.328 or 0.33    0.0792 or 0.11    0.2 or 0.099

**3.** 0.0057 or 0.006    0.4 or 0.0444    0.065 or 0.07

**Arrange each list in order from the smallest to the largest.**

**4.** 0.03, 0.33, 0.033, 0.303    0.082, 0.28, 0.8, 0.08

**5.** 0.106, 0.16, 0.061, 0.6    0.017, 0.2, 0.02, 0.007

**6.** 0.4, 0.405, 0.45, 0.045    0.04, 0.304, 0.32, 0.4

**7.** 0.0072, 0.07, 0.027, 0.02    0.2, 0.06, 0.0602, 0.026

# Rounding Decimals

To estimate answers to whole number problems, you can first round the numbers to numbers that end with zero. For example, to estimate the answer to $42 \times 68$, first round each number to the nearest ten. Then multiply the rounded numbers. $40 \times 70 = 2{,}800$

Rounding is also a useful skill for estimating answers to decimal problems.

To round a decimal, do the following steps.

**STEP 1** Underline the digit in the place you are rounding to.

**STEP 2** **a.** If the digit to the right of the underlined digit is *greater than or equal to 5*, add 1 to the underlined digit.

**b.** If the digit to the right of the underlined digit is *less than 5*, leave the underlined digit as is.

**STEP 3** Drop all the digits to the right of the underlined digit.

**EXAMPLE 1** **Round 2.38 to the nearest tenth.**

**STEP 1** Underline 3, the digit in the tenths place. $2.\underline{3}8 \rightarrow \mathbf{2.4}$

**STEP 2** The digit to the right of 3 is 8. Since 8 is greater than 5, add 1 to 3.

**STEP 3** Drop the digit to the right of 4.

**EXAMPLE 2** **Round 0.1627 to the nearest hundredth.**

**STEP 1** Underline 6, the digit in the hundredths place. $0.1\underline{6}27 \rightarrow \mathbf{0.16}$

**STEP 2** The digit to the right of 6 is 2. Since 2 is less than 5, leave 6 as is.

**STEP 3** Drop the digits to the right of 6.

When you round a decimal and add 1 to the digit 9, *two* decimal places will change. Study the next example carefully.

**EXAMPLE 3** **Round 3.00964 to the nearest thousandth.**

**STEP 1** Underline 9, the digit in the thousandths place. $3.00\underline{9}64 \rightarrow \mathbf{3.010}$

**STEP 2** The digit to the right of 9 is 6. Since 6 is greater than 5, add 1 to 9. $9 + 1 = 10$. Write 0 in the thousandths place and 1 in the hundredths place.

**STEP 3** Drop the digits to the right of the thousandths place.

Notice that the rounded number 3.010 ends in zero. You learned on page 68 that the zero to the right of 1 is unnecessary. However, the instruction in the example was to round the number to the nearest thousandth. The number 3.010 expresses thousandths.

**Round each decimal to the nearest place value given.**

| | | | | |
|---|---|---|---|---|
| **1.** tenth | 4.29 | 0.638 | 516.24 | 27.053 |
| **2.** hundredth | 0.582 | 12.487 | 0.0946 | 2.095 |
| **3.** thousandth | 23.4861 | 0.0537 | 5.4068 | 0.00349 |
| **4.** whole number (units or ones) | 17.26 | 1.89 | 356.541 | 199.8 |
| **5.** dollar (units or ones) | $14.39 | $9.78 | $1.66 | $347.09 |
| **6.** dime (tenth) | $2.79 | $63.52 | $0.98 | $5.34 |
| **7.** penny (hundredth) | $0.987 | $4.675 | $23.954 | $1.085 |

**Round 49.07583 to the nearest place value given.**

**8.** tenth　　hundredth　　thousandth　　ten-thousandth　　unit

**Write each fraction as a decimal. Then round the answer to the nearest *hundredth*.**

**9.** $\frac{5}{8} =$　　$\frac{2}{3} =$　　$\frac{3}{7} =$　　$\frac{7}{9} =$

**Write each fraction as a decimal. Then round the answer to the nearest *thousandth*.**

**10.** $\frac{1}{12} =$　　$\frac{5}{6} =$　　$\frac{3}{16} =$　　$\frac{2}{15} =$

# Adding Decimals

To add decimals, first line them up with *point under point.* Remember that any whole number is understood to have a decimal point at its right.

**EXAMPLE** 2.46 + 0.005 + 16 =

**STEP 1** Line up *point under point.* Notice the decimal point after the whole number 16.

**STEP 2** Add.

$$\begin{array}{r} 2.46\phantom{0} \\ 0.005 \\ +\ 16.\phantom{000} \\ \hline \mathbf{18.465} \end{array}$$

**Add.**

**1.** 0.8 + 0.047 + 0.36 = 4.9 + 17 + 3.28 =

**2.** 123 + 2.6 + 9.04 = 32.637 + 5 + 1.98 =

**3.** 9.043 + 0.27 + 15 = 8.04 + 26 + 31.263 =

**4.** 0.849 + 1.6 + 73 = 0.0097 + 2.8 + 16 =

**5.** 7.563 + 0.08 + 124.9 = 83.007 + 0.47 + 9.8 =

**6.** 12.3 + 0.908 + 6 + 4.25 = 0.0073 + 0.0196 + 0.08 + 0.4 =

**7.** 1.6 + 23 + 12.73 + 0.485 = 0.09 + 2.6 + 18 + 1.445 =

# Estimating Addition Answers

To estimate the answer to a decimal addition problem, first round the numbers in the problem. Then add the rounded numbers to estimate an answer. Compare the estimates in the next examples. The numbers in each estimate are rounded to a different place.

EXAMPLES **Estimate an answer to the problem 2.56 + 12.392 + 1.947 =**

Round each number to the nearest *tenth* and add. 2.6 + 12.4 + 1.9 = **16.9**

Round each number to the nearest *whole number* and add. 3 + 12 + 2 = **17**

Round each number to the *left-most* place that does not have a zero. Then add the rounded numbers. This is sometimes called **front-end rounding.** 3 + 10 + 2 = **15**

The exact answer is 2.56 + 12.392 + 1.947 = **16.899**

Notice that rounding to the nearest tenth gives the more accurate estimate, but the other two methods use easier numbers and will tell you whether an answer is reasonable.

**For problems 1 to 4, use** 16.34 + 52.908 + 132.15 =

**1.** Round to the nearest *tenth* and add.

**2.** Round to the nearest *unit* or *whole number* and add.

**3.** Use *front-end rounding* and add.

**4.** Find the exact answer.

**For problems 5 to 8, use** 0.586 + 1.437 + 2.184 =

**5.** Round to the nearest *hundredth* and add.

**6.** Round to the nearest *tenth* and add.

**7.** Round to the nearest *unit* or *whole number* and add.

**8.** Find the exact answer.

**For problems 9 to 12, use** 11.28 + 4.91 + 3.047 =

**9.** Round to the nearest *tenth* and add.

**10.** Round to the nearest *unit* or *whole number* and add.

**11.** Use *front-end rounding* and add.

**12.** Find the exact answer.

# Applying Your Addition Skills

**Solve and write the correct label, such as inches or pounds, next to each answer. Use front-end rounding to estimate each answer.**

1. The average monthly rainfall in New York City is 2.96 inches in June, 3.69 inches in July, and 4.01 inches in August. What is the total rainfall for these three months?

2. For his job as a truck driver, Jack drove a four-day route. He drove 278.5 miles on Thursday, 243.7 miles on Friday, 386 miles on Saturday, and 291.8 miles on Sunday. What total distance did he drive on this route?

3. Aldo welded together pieces of pipe that were 25.6 inches long, 19.8 inches long, and 31.5 inches long. How long was the pipe made of the three welded pieces?

4. Dorothy bought 2.6 pounds of beef, 1.75 pounds of cheese, 3 pounds of chicken, and 2.35 pounds of fish. What was the total weight she had to carry?

5. The distance from Middletown to Springfield is 72.6 miles. The distance from Springfield to Centerville is 48.9 miles. What is the distance from Middletown to Centerville by driving through Springfield?

6. Maceo is a licensed practical nurse. He must alert the doctor if a patient's temperature is 4.5° above normal. A patient's normal temperature is 98.6°. At what temperature must he send an alert?

7. Find the total weight of three cartons that weigh 4.2 kilograms, 2.37 kilograms, and 1.45 kilograms.

8. In a recent year, the population of Mexico was 111.2 million people. Experts think that in 2050 Mexico will have 41.9 million more people. If the experts are correct, what will be the population of Mexico in 2050?

# Subtracting Decimals

To subtract decimals, put the larger number on top and line up the decimal points. Use zeros to give each decimal the same number of decimal places. Then subtract and bring down the decimal point.

EXAMPLE $15.2 - 0.184 =$

**STEP 1** Put 15.2 on top and line up the decimal points.

**STEP 2** Write two zeros to the right of 15.2 to give 15.2 the same number of decimal places as 0.184.

**STEP 3** Subtract and bring down the decimal point.

$$\begin{array}{r} 15.200 \\ -\ 0.184 \\ \hline \mathbf{15.016} \end{array}$$

**Subtract.**

**1.** $4.2 - 3.76 =$ $0.804 - 0.1673 =$ $3.2 - 2.68 =$

**2.** $0.2 - 0.078 =$ $60.4 - 6.04 =$ $89.3 - 0.766 =$

**3.** $12 - 0.936 =$ $1 - 0.047 =$ $13 - 0.932 =$

**4.** $8.4 - 0.631 =$ $0.07 - 0.002 =$ $5 - 2.493 =$

**5.** $96 - 7.485 =$ $3.2 - 0.1986 =$ $0.47 - 0.3992 =$

**6.** $1.3 - 1.064 =$ $0.81 - 0.793 =$ $2 - 0.807 =$

**7.** $0.08 - 0.0156 =$ $1.4 - 0.978 =$ $0.6 - 0.059 =$

# Estimating Subtraction Answers

To estimate the answer to a decimal subtraction problem, first round the numbers in the problem. Then subtract the rounded numbers to estimate an answer. Compare the estimates in the next examples. The numbers in each estimate are rounded to a different place.

EXAMPLES Estimate an answer to the problem 12.54 − 7.384 =

Round each number to the nearest *tenth* and subtract. 12.5 − 7.4 = **5.1**

Round each number to the nearest *whole number* and subtract. 13 − 7 = **6**

Use *front-end* rounding. Then subtract the rounded numbers. 10 − 7 = **3**

The exact answer is 12.56 − 7.4 = **5.16.**

Notice that rounding to the nearest tenth gives the more accurate estimate. Rounding to the nearest whole number gives a reasonable estimate, but front-end rounding is less accurate.

**For problems 1 to 4, use** 9.47 − 0.926 =

1. Round to the nearest *tenth* and subtract.
2. Round to the nearest *unit* or *whole number* and subtract.
3. Use *front-end rounding* and subtract.
4. Find the exact answer.

**For problems 5 to 9, use** 42.713 − 11.085 =

5. Round to the nearest *hundredth* and subtract.
6. Round to the nearest *tenth* and subtract.
7. Round to the nearest *unit* or *whole number* and subtract.
8. Round to the nearest *ten* and subtract.
9. Find the exact answer.

**For problems 10 to 14, use** 217.64 − 59.372 =

10. Round to the nearest *tenth* and subtract.
11. Round to the nearest *unit* or *whole number* and subtract.
12. Round to the nearest *ten* and subtract.
13. Use *front-end rounding* and subtract.
14. Find the exact answer.

# Applying Your Subtraction Skills

Subtraction problems ask you to figure out what is *left* after something is taken away or they ask you to compare quantities. Remember to put the larger number on top.

**For problems 1 to 8, solve and write the correct label, such as years or meters, next to each answer. Then round each number to the nearest whole number, and use the rounded numbers to estimate the answers.**

1. In a recent year, the average man in the United States lived for 75.1 years. The average woman lived for 80.2 years. The average woman's life expectancy was how much more than the average man's life expectancy?

2. The area of Philadelphia is 135.1 square miles. The area of Pittsburgh is 55.6 square miles. The area of Philadelphia is how much greater than the area of Pittsburgh?

3. From a board that was 3 meters long, Colin cut off a piece 1.67 meters long. How long was the piece that was left?

4. The distance from the Smiths' house to the cabin where they plan to spend their vacation is 91.4 miles on the interstate highway or 104.2 miles on more scenic back roads. The drive on the scenic road is how much longer than the drive on the interstate highway?

5. When Sam began driving Monday morning, the mileage gauge on his moving van read 16,428.7 miles. When he stopped driving Monday night, it read 16,831.4 miles. How many miles did Sam drive that day?

6. The metal platform of a bridge is 124.2 meters long in the summer. In the winter it shrinks by 1.05 meters. How long is the bridge platform in the winter?

7. One year there were 99.7 million cable television subscribers in the United States. The next year there were 103 million subscribers. By how many did the number of subscribers increase?

8. The price of 1 pound of steak rose from $6.99 to $7.34. By how much did the price of 1 pound of beef increase? Estimate by rounding the prices to the nearest $0.10.

**Read the next problems carefully to decide whether you should add or subtract.**

**9.** Andy works in the shipping department of a publisher. He has to send a customer three textbooks that weigh 0.8 kilogram, 0.44 kilogram, and 0.65 kilogram. What is the total weight of the three books?

**10.** On the carton that Andy plans to use for shipping the books, there is a label that says the carton can hold a maximum of 2 kilograms. How much over or under the maximum is the total weight of the three books?

**11.** One summer the average price of a gallon of gasoline in Cleveland was \$2.709 while the average price of gasoline in New York City was \$3.028. To the nearest penny, how much more did a gallon of gasoline cost in New York City than in Cleveland?

**12.** At the same time as the problem above, the average price of a gallon of gasoline in Honolulu was \$3.523. To the nearest penny, how much more was the price of a gallon of gasoline in Honolulu than in Cleveland?

**13.** According to the U.S. Census Bureau, there are currently 40.2 million Americans who are 65 and older. The projection for 2050 is that there will be another 48.3 million Americans in this age category. If the projection is accurate, how many Americans 65 and older will there be in 2050?

**14.** The normal September rainfall for Tucson is 1.45 inches. The normal rainfall for Miami in September is 8.38 inches. To the nearest tenth of an inch, how much more rain does Miami get than Tucson?

**15.** Miami gets the least rain in January when the normal rainfall is 1.88 inches. Use the information in the last problem to tell how much more rain Miami gets in September than it gets in January.

**16.** A nationally read newspaper sells 1.6 million copies each day. Of these copies, 0.4 million are distributed by train, and 0.15 million are distributed by plane. The rest are distributed by truck. How many copies of the newspaper are distributed by truck each day?

# Multiplying Decimals

To multiply decimals, multiply the two numbers the same way you would whole numbers. Then count the number of decimal places in both numbers you are multiplying. Decimal places are to the right of the decimal point. Put the total number of places in your answer.

**EXAMPLE 1** $4.36 \times 2 =$

| | |
|---|---|
| two decimal places | 4.36 |
| no decimal places | $\times$ 2 |
| two decimal places | **8.72** |

**Multiply.**

**1.** $3.8 \times 4$ $\quad 0.92 \times 9$ $\quad 6.7 \times 6$ $\quad 5.3 \times 8$ $\quad 0.84 \times 7$

**2.** $41 \times 0.03$ $\quad 78 \times 0.5$ $\quad 59 \times 0.09$ $\quad 86 \times 0.4$ $\quad 19 \times 0.06$

**3.** $34.7 \times 8$ $\quad 2.89 \times 7$ $\quad 0.551 \times 6$ $\quad 60.3 \times 9$ $\quad 7.18 \times 4$

**4.** $906 \times 0.07$ $\quad 504 \times 0.002$ $\quad 783 \times 0.8$ $\quad 652 \times 0.06$ $\quad 467 \times 0.003$

**5.** $2.85 \times 50$ $\quad 0.693 \times 40$ $\quad 44.7 \times 30$ $\quad 8.01 \times 70$ $\quad 9.17 \times 60$

You may need to add zeros in front of your answer to have enough decimal places in the final answer.

EXAMPLE 2 $0.06 \times 0.4 =$

| | |
|---|---|
| two decimal places | 0.06 |
| one decimal place | $\times$ 0.4 |
| *needs one zero to make three places* | _24 |
| three decimal places | **0.024** |

**6.** $0.09 \times 0.6$    $0.05 \times 0.7$    $0.004 \times 0.3$    $0.08 \times 0.04$    $0.002 \times 0.8$

**7.** $5.6 \times 0.2$    $0.73 \times 0.08$    $9.2 \times 0.7$    $0.087 \times 0.4$    $3.3 \times 0.06$

**8.** $41.8 \times 0.7$    $3.90 \times 0.08$    $0.516 \times 0.5$    $73.8 \times 0.06$    $3.47 \times 0.4$

**9.** $4.5 \times 2.6$    $0.48 \times 5.2$    $0.92 \times 8.9$    $3.4 \times 0.71$    $0.39 \times 0.56$

**10.** $0.27 \times 1.8$    $5.6 \times 0.39$    $4.9 \times 5.4$    $0.83 \times 0.17$    $9.2 \times 0.66$

**Rewrite each problem and multiply.**

**11.** $2.8 \times 4.3 =$ $\quad$ $5.6 \times 0.82 =$ $\quad$ $0.72 \times 5.7 =$

**Note:** Parentheses with no operation signs between them also mean to multiply.

**12.** $(0.81)(0.69) =$ $\quad$ $(0.94)(1.8) =$ $\quad$ $(30.5)(0.27) =$

**13.** $7.4 \times 6.6 =$ $\quad$ $61.8 \times 4.8 =$ $\quad$ $0.514 \times 0.91 =$

**14.** $(9.06)(7.3) =$ $\quad$ $(45.21)(5.6) =$ $\quad$ $(3.748)(73) =$

**15.** $206.9 \times 0.28 =$ $\quad$ $0.7488 \times 4.9 =$ $\quad$ $34.7 \times 209 =$

**16.** $(6.82)(5.31) =$ $\quad$ $(0.925)(78.8) =$ $\quad$ $(8.43)(0.409) =$

**17.** $17.32 \times 0.16 =$ $\quad$ $0.0648 \times 2.3 =$ $\quad$ $913.2 \times 0.049 =$

**18.** $(40.21)(20.8) =$ $\quad$ $(0.2789)(0.17) =$ $\quad$ $(3.845)(29.2) =$

**19.** $5{,}183.6 \times 0.0016 =$ $\quad$ $303.003 \times 56.8 =$ $\quad$ $7.05 \times 0.408 =$

# Multiplying Decimals by 10, 100, and 1,000

To multiply a decimal by 10, move the decimal point *one place to the right.*

EXAMPLE 1 $0.26 \times 10 =$

Move the decimal point one place to the right.
$0.26 \times 10 = 2.6 =$ **2.6**

To multiply a decimal by 100, move the decimal point *two places to the right.* You may have to add zeros to get enough places.

EXAMPLE 2 $3.7 \times 100 =$

Move the decimal point two places to the right.
$3.7 \times 100 = 3.70 =$ **370**

To multiply a decimal by 1,000, move the decimal point *three places to the right.*

EXAMPLE 3 $1.4 \times 1{,}000 =$

Move the decimal point three places to the right.
$1.4 \times 1{,}000 = 1.400 =$ **1,400**

**Multiply.**

**1.** $0.8 \times 10 =$ $0.09 \times 10 =$ $3.64 \times 10 =$ $0.721 \times 10 =$

**2.** $0.03 \times 100 =$ $0.275 \times 100 =$ $8.9 \times 100 =$ $0.863 \times 100 =$

**3.** $0.9 \times 1{,}000 =$ $2.36 \times 1{,}000 =$ $0.475 \times 1{,}000 =$ $1.6 \times 1{,}000 =$

**4.** $0.34 \times 10 =$ $1.24 \times 100 =$ $3.85 \times 1{,}000 =$ $0.06 \times 1{,}000 =$

**5.** $\$1.25 \times 10 =$ $\$0.60 \times 100 =$ $\$2.25 \times 1{,}000 =$ $\$0.03 \times 1{,}000 =$

# Estimating Multiplication Answers

Remember that a decimal is a kind of fraction. When you find *a fraction of* a number, you find *a part of* that number. For example, $\frac{1}{3}$ of 18 is 6. In some situations, you can use general rules to get a sense if your answer is correct.

When you multiply a number greater than 0 by a decimal (a number with no digits in the whole-number places), the answer is always *less than* that number.

EXAMPLES $0.8 \times 6 = 4.8$ $0.3 \times 1.2 = 0.36$ $0.135 \times 100 = 13.5$

When you multiply a number greater than 0 by a mixed decimal, the answer is always *greater than* that number.

EXAMPLES $1.5 \times 6 = 9$ $2.5 \times 4.3 = 10.75$ $1.9 \times 50 = 95$

**In each box write the symbol <, >, or = to make the statement true.**

1. $4.2 \times 3.8$ ☐ $3.8$ $0.6 \times 7.8$ ☐ $7.8$ $1 \times 0.025$ ☐ $0.025$

2. $0.14 \times 407$ ☐ $407$ $10.5 \times 2.39$ ☐ $2.39$ $5 \times 0.88$ ☐ $0.88$

To estimate an answer to a decimal multiplication problem with mixed decimals, you can use front-end rounding. Round each number to the *left-most* place that does not have a zero. Then multiply the rounded numbers.

EXAMPLE Use front-end rounding to estimate the answer to $2.84 \times 0.72$.

STEP 1 Round 2.84 to the nearest unit, and round 0.72 to the nearest tenth. $2.84 \times 0.72 \approx 3 \times 0.7$

STEP 2 Multiply the rounded numbers. $3 \times 0.7 = \mathbf{2.1}$

The exact answer to the problem is 2.0448.

**Use front-end rounding to estimate an answer to each problem. Then find the exact answer.**

3. $4.7 \times 0.3 =$ $0.83 \times 0.09 =$ $6.4 \times 0.8 =$

4. $7.9 \times 0.06 =$ $0.58 \times 0.3 =$ $1.82 \times 0.7 =$

5. $5.2 \times 3.7 =$ $0.62 \times 4.1 =$ $0.74 \times 0.53 =$

6. $0.74 \times 0.38 =$ $9.1 \times 0.67 =$ $0.81 \times 0.44 =$

7. $1.907 \times 5.2 =$ $42.7 \times 0.41 =$ $0.681 \times 9.2 =$

# Applying Your Multiplication Skills

In each problem pay close attention to the language that tells you to multiply. In most cases you will be given information about *one thing*, and you will be asked to find information about *several things*.

**For problems 1 to 9, solve each problem and write the correct label beside each answer. Then, to decide whether each answer is reasonable, use front-end rounding to estimate the answer.**

1. Alberto makes $19.20 an hour at his job repairing office equipment. If he worked 7.5 hours on Monday, how much did he earn that day?

2. An airplane flew at an average speed of 386 miles per hour for 2.75 hours. How far did the plane fly?

3. The average household in Middleville has 2.3 people. There are approximately 9,800 households in Middleville. What is the population of the town?

4. What is the cost of 0.85 pound of imported cheese at $7.90 per pound?

5. Matt works in the shipping department of a printing company. He has to ship 1,150 copies of a new telephone book. If each book weighs 0.45 pounds, what is the total weight of the shipment?

6. Irma works in a medical testing lab. She has to fill 18 bottles with saline solution. Each bottle holds 0.75 liter. What is the total volume of the saline solution she will need?

7. It costs $0.84 per hour to run the lights in a small movie theater. How much does it cost to run the lights for 11.5 hours?

8. A certain electric cable costs $18.75 a meter. What is the cost of 21.5 meters of the cable?

9. One cubic foot of water weighs 62.5 pounds. A cooling tower holds 58 cubic feet of water. What is the weight of the water in the tower?

**Read the next problems carefully to decide which operation to use.**

**10.** The gasoline tank in Deborah's car holds 13.2 gallons. When the gasoline gauge says that the tank is $\frac{1}{4}$ full, how many gallons of gas does the tank have?

**11.** In the last problem, when Deborah's tank is $\frac{1}{4}$ full, how many gallons of gas does she have to buy to fill the tank?

**12.** Andrea is a receptionist in a law office. She answers the phone 12.8 times per hour. If she sits at the telephone 6.5 hours a day, how many calls does she usually answer in one day?

**13.** When Andrea, in the last problem, is not at the receptionist's desk, she files documents. During a normal 8-hour day, how many hours does she spend filing?

**14.** One pound is equal to 0.45 kilogram. Find the weight in kilograms of a person who weighs 160 pounds.

**15.** Mike pays a commercial rate for electricity at his electronics repair shop. One month he used 883 kilowatt-hours of electricity at the rate of $0.153 per kilowatt-hour. Find the cost for his electrical use that month.

**16.** Mike also had to pay a monthly fee of $7.82 and tax of $8.11 to the electric company. What was Mike's total bill for electricity that month?

Jessie works in a large warehouse. He has to pack welding rods in boxes that will each weigh 6.5 kilograms. Use this information to answer questions 17 to 20.

**17.** The Standard Steel Construction Company placed an order for 240 boxes of welding rods. Find the total weight of the order.

**18.** The freight elevator in the warehouse can hold a maximum of 1,650 kilograms. With the entire order from the last problem loaded on the elevator, how much additional weight can the elevator carry?

**19.** One kilogram is approximately 2.2 pounds. When the elevator is loaded with this order, how much additional weight, in pounds, can the elevator carry?

**20.** Jessie weighs 235 pounds. Should he ride in the elevator with the load? Why?

# Dividing Decimals by Whole Numbers

To divide a decimal by a whole number, bring the decimal point up in the answer directly above its position in the problem. Then divide as you would whole numbers.

EXAMPLE 1 $9.32 \div 4 =$ $\quad 4\overline{)9.32}$ with quotient **2.33**

Remember that there are several ways to write a division problem. In each example below, 9.32 is the **dividend,** 4 is the **divisor,** and 2.33 is the answer or the **quotient.**

$4\overline{)9.32}$ with quotient 2.33 $\qquad 9.32 \div 4 = 2.33 \qquad \frac{9.32}{4} = 2.33 \qquad 9.32/4 = 2.33$

You may have to put a zero in the quotient as a decimal placeholder.

EXAMPLE 2 $0.222 \div 6 =$

STEP 1 Bring the decimal point up in the quotient and put a zero in the tenths place. $\quad 6\overline{)0.222}$ with quotient **0.037**

STEP 2 Divide.

**Divide.**

**1.** $6\overline{)13.8}$ $\qquad 9\overline{)70.2}$ $\qquad 4\overline{)0.384}$ $\qquad 7\overline{)57.75}$

**2.** $8\overline{)0.192}$ $\qquad 3\overline{)148.8}$ $\qquad 5\overline{)19.45}$ $\qquad 4\overline{)2.524}$

**Rewrite each problem and divide.**

**3.** $76.8 \div 16 =$ $\qquad 7.56 \div 21 =$ $\qquad 1.52 \div 19 =$ $\qquad 216.6 \div 38 =$

**4.** $\frac{9.516}{52} =$ $\qquad \frac{1{,}565.2}{43} =$ $\qquad \frac{464.31}{77} =$ $\qquad \frac{33.605}{65} =$

# Dividing Decimals by Decimals

To divide a decimal by a decimal, first change the problem to a new problem with a whole number divisor. Then move the point in the dividend the same number of places that you moved the point in the divisor. Study Example 1 carefully.

EXAMPLE 1 4.374 ÷ 0.03 =

**STEP 1** Move the decimal point in the divisor to the right as far as it will go.

$0.03\overline{)4.374}$

**STEP 2** Move the decimal point in the dividend the same number of places that you moved the decimal point in the divisor.

$0.03\overline{)4.37\,4}$

**STEP 3** Bring the decimal point up above its new position and divide.

$\begin{array}{r} \mathbf{145.8} \\ 0.03\overline{)4.37\,4} \end{array}$

When you move the decimal point in both the divisor and the dividend, you get a new problem that has the same answer. This is easier to understand with whole numbers.

Think about the problem 12 ÷ 4 = 3. If you multiply both the divisor and the dividend by 10, you get the problem 120 ÷ 40 = 3, which has the same answer.

**Divide.**

**1.** $0.8\overline{)7.68}$ $0.7\overline{)26.6}$ $0.6\overline{)0.42}$ $0.4\overline{)2.76}$

**2.** $4.1\overline{)1.148}$ $9.2\overline{)128.8}$ $5.3\overline{)4.399}$ $3.2\overline{)0.288}$

**Rewrite each problem and divide.**

**3.** 0.882 ÷ 0.06 = 0.5404 ÷ 0.28 = 453.75 ÷ 0.75 = 6.656 ÷ 0.52 =

**4.** 1.17 ÷ 0.9 = 0.888 ÷ 3.7 = 5.92 ÷ 0.16 = 1.334 ÷ 0.46 =

Sometimes you need to add zeros in a problem in order to have enough places to move the decimal point.

**EXAMPLE 2** $4.8 \div 0.08 =$

**STEP 1** Move the decimal point in the divisor two places. $0.08\overline{)4.8}$

**STEP 2** To move the decimal point two places to the right in the dividend, add one zero. $0.08\overline{)4.80}$

**STEP 3** Bring the decimal point up and divide. $\overset{\mathbf{60.}}{0.08\overline{)4.80}}$

**Divide.**

**5.** $0.007\overline{)5.32}$ $0.009\overline{)4.32}$ $0.003\overline{)8.1}$ $0.008\overline{)77.12}$

**6.** $0.016\overline{)212.8}$ $0.025\overline{)0.1}$ $0.091\overline{)65.52}$ $0.68\overline{)57.8}$

**Rewrite each problem and divide.**

**7.** $0.78 \div 0.0024 =$ $42.84 \div 0.0006 =$ $1.683 \div 0.0018 =$ $15.184 \div 0.0073 =$

**8.** $\frac{156.8}{0.32} =$ $\frac{1{,}381.5}{0.45} =$ $\frac{4.48}{0.008} =$ $\frac{558.6}{0.06} =$

**9.** $\frac{0.522}{8.7} =$ $\frac{0.3933}{0.19} =$ $\frac{145.44}{3.6} =$ $\frac{0.11648}{0.64} =$

# Dividing Whole Numbers by Decimals

To divide a whole number by a decimal, put a decimal point after the whole number and add zeros in order to move the decimal point enough places. Remember that a whole number is understood to have a decimal point at its right.

EXAMPLE $35 \div 0.007 =$

**STEP 1** Move the decimal point in the divisor three places to the right. $0.007\overline{)35}$

**STEP 2** Put a decimal point to the right of the whole number. Then write three zeros and move the decimal point three places to the right. $0.007\overline{)35.000}$

**STEP 3** Bring the decimal point up and divide. $\begin{array}{r}\textbf{5 000.}\\0.007\overline{)35.000}\end{array}$

**Divide.**

**1.** $0.4\overline{)28}$ $0.9\overline{)360}$ $0.03\overline{)111}$ $0.08\overline{)512}$

**2.** $0.007\overline{)63}$ $0.0005\overline{)4}$ $0.06\overline{)234}$ $0.3\overline{)1{,}884}$

**Rewrite each problem and divide.**

**3.** $\frac{18}{0.09} =$ $\frac{12}{0.003} =$ $\frac{40}{0.02} =$ $\frac{7}{0.25} =$

**4.** $\frac{552}{1.2} =$ $\frac{2{,}592}{0.36} =$ $\frac{2{,}135}{4.27} =$ $\frac{21{,}546}{51.3} =$

**5.** $\frac{1{,}178}{0.019} =$ $\frac{37{,}440}{0.48} =$ $\frac{33{,}040}{5.6} =$ $\frac{3{,}237}{0.039} =$

# Dividing Decimals by 10, 100, and 1,000

To divide a decimal by 10, move the decimal point *one place to the left.*

EXAMPLE 1 $7.2 \div 10 =$

Move the decimal point one place to the left.

$7.2 \div 10 = .72 =$ **0.72**

To divide a decimal by 100, move the decimal point *two places to the left.* You may have to add zeros to get enough places.

EXAMPLE 2 $3.64 \div 100 =$

Move the decimal point two places to the left.

$3.64 \div 100 = .0364 =$ **0.0364**

To divide a decimal by 1,000, move the decimal point *three places to the left.*

EXAMPLE 3 $25.3 \div 1{,}000 =$

Move the decimal point three places to the left.

$25.3 \div 1{,}000 = .0253 =$ **0.0253**

**Divide.**

**1.** $0.9 \div 10 =$ $36 \div 10 =$ $27.3 \div 10 =$ $0.04 \div 10 =$

**2.** $14.2 \div 100 =$ $1.3 \div 100 =$ $728 \div 100 =$ $0.6 \div 100 =$

**3.** $37.5 \div 1{,}000 =$ $1.8 \div 1{,}000 =$ $2 \div 1{,}000 =$ $428 \div 1{,}000 =$

**4.** $13.45 \div 10 =$ $0.32 \div 100 =$ $6{,}954 \div 1{,}000 =$ $15.8 \div 1{,}000 =$

**5.** $\$2.90 \div 10 =$ $\$650 \div 100 =$ $\$540 \div 1{,}000 =$ $\$20 \div 1{,}000 =$

# Dividing to Fixed-Place Accuracy

Adding zeros to a division of decimals problem does not always result in a problem that divides evenly. If the numbers in a division problem include tenths, an answer rounded to the nearest tenth is often accurate enough.

To get a division answer that is accurate to the nearest *tenth*, divide to the hundredths place and round the answer to the nearest tenth. This is called dividing to **fixed-place accuracy.** In this case, the fixed place is tenths.

EXAMPLE Find the answer to 4.3 ÷ 0.7 to the nearest tenth.

**STEP 1** Move the decimal point in both the divisor and the dividend one place to the right.

$0.7\overline{)4.3}$

**STEP 2** Add two zeros to get one more place beyond the tenths place.

$0.7\overline{)4.300}$

**STEP 3** Divide to the hundredths place and round the answer to the nearest tenth.

$0.7\overline{)4.300}$ = 6.14 → **6.1**

If you solve the example on a calculator, the calculator display will show the answer 6.1428571, which, rounded to the nearest tenth is **6.1.**

**Find the answer to each problem to the nearest tenth.**

**1.** 2 ÷ 1.8 = 4.1 ÷ 0.36 = 0.4 ÷ 0.12 =

**2.** 1.2 ÷ 9 = 0.17 ÷ 0.6 = 5 ÷ 0.7 =

**3.** 37.4 ÷ 12 = 0.45 ÷ 2.1 = 18 ÷ 0.13 =

**4.** 49.5 ÷ 36 = 10.3 ÷ 0.35 = 204 ÷ 9.2 =

**5.** 12.3 ÷ 7 = 72.6 ÷ 50.3 = 63 ÷ 0.8 =

# Estimating Division Answers

Remember that in the problem $6 \div 0.5 = 12$, 6 is the dividend, 0.5 is the divisor, and 12 is the quotient. The problem states that "6 divided by 5 tenths is equal to 12." In some situations, you can use general rules to get a sense if your answer is correct.

**When a number greater than 0 is divided by a decimal, the quotient is always *greater than* the dividend.**

When a whole number or a mixed decimal is divided by a decimal, the answer is always *greater than* the whole number or the mixed decimal.

EXAMPLES $6 \div 0.5 = 12$ $\quad 2.5 \div 0.25 = 10$ $\quad 18 \div 4.5 = 4$

**When a number greater than 0 is divided by a larger number, the quotient is always *a decimal*.**

When a decimal with no whole number is divided by a whole number or a mixed number, the answer is always a *smaller* decimal.

EXAMPLES $0.3 \div 6 = 0.05$ $\quad 0.75 \div 1.5 = 0.5$ $\quad 1.64 \div 8 = 0.205$

**In each box, write the symbol < (less than) or > (greater than) to make the statement true.**

1. $6.5 \div 2$ ☐ $6.5$ $\quad 1.44 \div 0.6$ ☐ $1.44$ $\quad 3 \div 0.5$ ☐ $3$

2. $0.9 \div 0.15$ ☐ $0.9$ $\quad 2.4 \div 4.8$ ☐ $2.4$ $\quad 1 \div 0.48$ ☐ $1$

If the situations above do not apply, you can still get an approximation to check your answer. Estimating decimal division answers can be almost as difficult as solving the problems. However, front-end rounding makes the numbers in a problem easier. Use front-end rounding to estimate answers.

EXAMPLE Estimate an answer to the problem $3.6 \div 2.4$.

**STEP 1** Round both numbers to the nearest whole number. $3.6 \rightarrow 4$ and $2.4 \rightarrow 2$

**STEP 2** Divide the rounded numbers. $4 \div 2 = \mathbf{2}$

The exact answer is $3.6 \div 2.4 = 1.5$

**Solve each problem. Then, to see whether your answer is reasonable, use front-end rounding to estimate each answer.**

3. $2.76 \div 0.6 =$ $\quad 0.75 \div 5 =$ $\quad 9.3 \div 1.5 =$

4. $0.64 \div 3.2 =$ $\quad 27.88 \div 8.2 =$ $\quad 0.33 \div 0.44 =$

5. $0.36 \div 1.5 =$ $\quad 11.4 \div 7.6 =$ $\quad 26.4 \div 1.6 =$

6. $58.8 \div 4.2 =$ $\quad 10.8 \div 1.35 =$ $\quad 0.35 \div 1.4 =$

# Applying Your Division Skills

Pay close attention to the language that tells you to divide.

A problem may give information about several things and ask you to find information for one of these things.

You may also be asked to find an average which is a total divided by the number of items that make up the total.

**For problems 1 to 6, solve each problem and write the correct label, such as pounds or miles, next to each answer. Then, to see whether your answers are reasonable, estimate each answer. In parentheses after each problem are suggestions on how to make each estimate.**

1. Together, the four members of the Rosa family weigh 508.8 pounds. What is the average weight of one member of the family? (Round the total weight to the nearest hundred.)

2. Mark drove a total of 1,247.5 miles in 5 days. What was the average number of miles he drove each day? (Round the distance to the nearest hundred.)

3. The cost of 3.5 yards of material is $33.60. Find the cost of 1 yard of the material. (Use front-end rounding.)

4. Margie made $477.40 last week. She makes $12.40 an hour. How many hours did she work last week? (Use front-end rounding.)

5. A developer plans to divide a 41.5-acre piece of land into five equal parcels. Find the size of each parcel. (Use front-end rounding.)

6. If 2.3 pounds of vegetables cost $3.68, what is the price of 1 pound of vegetables? (Use front-end rounding.)

7. There are 1.6 kilometers in a mile. How many miles are there in 36.8 kilometers?

8. There are 2.54 centimeters in 1 inch. How many inches are there in 45.72 centimeters?

9. In baseball, a batting average is the number of hits a player gets divided by the number of times he is at bat. Batting averages are measured in *thousandths*. Find the batting average of a player who gets 17 hits in 52 times at bat. (**Hint:** Divide to the ten-thousandths place and round your answer to the nearest thousandth.)

10. Find the batting average of another baseball player who got 14 hits in 45 times at bat.

11. In basketball, a scoring average is the number of points a player gets divided by the number of games he plays. Basketball scoring averages are measured in *tenths*. One season, Kobe Bryant scored 2,832 points in 80 games. Find his scoring average for the season. (**Hint:** Divide to the hundredths place and round your answer to the nearest tenth.)

12. Find the scoring average of another basketball player who scored 1,710 points in 82 games.

13. Camille agreed to pay $49.50 a month for new living-room furniture. How many months will it take her to pay the $990 that she owes?

14. In one week, a factory produced 700 chairs. There are 18 workers in the factory. To the nearest whole number, how many chairs were produced per worker that week?

15. Joshua works at a publishing company. He typed a 238-page document in 3.5 hours. What was the average number of pages that he typed in one hour?

# Cumulative Review

This review covers the material you have studied so far in this book. When you finish, check your answers at the back of the book.

**1.** $3\frac{1}{8} + 5\frac{2}{3} =$

**2.** $10\frac{1}{2} - 6\frac{5}{9} =$

**3.** $5\frac{1}{3} \times 2\frac{1}{2} =$

**4.** $2\frac{1}{2} \div \frac{5}{8} =$

**5.** Write fourteen ten-thousandths as a decimal.

**6.** Write 0.016 as a fraction and reduce.

**7.** Write $\frac{7}{16}$ as a decimal.

**8.** Which is larger, 0.0013 or 0.02?

**9.** Rewrite the following list in order from smallest to largest:

0.03, 0.031, 0.1, 0.013

**10.** What is 12.384 rounded to the nearest tenth?

**11.** $0.056 + 2.3 + 19 =$

**12.** $8 + 1.0573 + 1.64 =$

**13.** Janet bought 2.6 pounds of beef, 1.3 pounds of cheese, 2.45 pounds of chicken, and 5 pounds of sugar. What was the total weight she had to carry?

**14.** $12.3 - 0.097 =$

**15.** $3 - 0.654 =$

**16.** In polishing a piece of pipe that was 2 inches thick, 0.016 inch of metal was worn away. What was the thickness of the pipe when it was polished?

**17.** $63.2 \times 0.045 =$

**18.** $20.8 \times 100 =$

**19.** In the problem 91.12 × 7.8, round 91.12 to the nearest *ten* and 7.8 to the nearest *unit.* Then multiply.

**20.** One inch is equal to 2.54 centimeters. How many centimeters are there in 6.5 inches? Round your answer to the nearest tenth.

**21.** At $1.80 a meter, how much do 3.75 meters of lumber cost?

**22.** 69.53 ÷ 17 =

**23.** 2.016 ÷ 0.28 =

**24.** 14.1 ÷ 2.35 =

**25.** 9 ÷ 0.036 =

**26.** 5.13 ÷ 1,000 =

**27.** A train traveled 145 miles in 4.2 hours. To the nearest tenth, what was the average speed of the train in miles per hour?

**28.** Manny made $102.05 for 6.5 hours of work. How much did he make per hour?

## CUMULATIVE REVIEW CHART

If you missed more than one problem on any group below, review the practice pages for those problems. Then redo the problems you got wrong before going on to the Percent Skills Inventory. If you had a passing score, redo any problem you missed and begin the Percent Skills Inventory on the next page.

| Problem Numbers | Skill Area | Practice Pages |
|---|---|---|
| 1, 2, 3, 4 | using fractions | 11–59 |
| 5, 6, 7, 8, 9, 10 | understanding decimals | 64–73 |
| 11, 12, 13 | adding decimals | 74–76 |
| 14, 15, 16 | subtracting decimals | 77–80 |
| 17, 18, 19, 20, 21 | multiplying decimals | 81–87 |
| 22, 23, 24, 25, 26, 27, 28 | dividing decimals | 88–96 |

# Percent Skills Inventory

This inventory will let you know whether you need to work through the percent section of this book. Do all the problems that you can. Work carefully and check your answers, but do not use outside help. Correct answers are listed by page number at the back of the book.

**1.** Write 0.017 as a percent.

**2.** Write 4% as a decimal.

**3.** Write $\frac{5}{8}$ as a percent.

**4.** Write 12% as a proper fraction.

**5.** Write $8\frac{1}{3}$% as a proper fraction.

**6.** Find 15% of 80.

**7.** Find 4.6% of 250.

**8.** Find $37\frac{1}{2}$% of 96.

**9.** The sales tax in Tony's state is 8%. How much tax does he pay for a baseball cap that costs $10.50?

**10.** The Olsons spend $33\frac{1}{3}$% of their take-home pay on mortgage payments. If their take-home pay is $3,378 a month, what is their monthly mortgage payment?

**11.** Niko works in construction. He was making $12.50 an hour. His supervisor is giving him a 6% raise. What will be his new hourly wage?

**12.** 15 is what percent of 45?

**13.** 32 is what percent of 400?

**14.** On a loan of $500, Alfonso had to pay $55 in interest. The interest represents what percent of the loan?

**15.** There were 36 people at a meeting of a tenants' organization. The total number of members in the organization is 54. What percent of the members attended the meeting?

**16.** The number of workers at Eldridge Electronics increased from 65 to 91 last year. By what percent did the number of workers increase?

**17.** 75% of what number is 60?

**18.** $62\frac{1}{2}\%$ of what number is 35?

**19.** The Morris family spends $1,270 each month to rent their two-bedroom apartment. If rent represents 25% of their monthly income, what is their monthly income?

**20.** Chuck spent $1.59 in tax for a new shirt. The tax rate in his state is 6%. What was the price of the shirt?

**21.** Kate works for a company that investigates consumer satisfaction. Her company found that only 35% of the people they telephoned were happy with their cable television service. If 511 people said they were satisfied with their cable service, how many people did Kate's company call?

## PERCENT SKILLS INVENTORY CHART

If you missed more than one problem on any group below, work through the practice pages for that group. Then redo the problems you got wrong on the Percent Skills Inventory. If you had a passing score on all four groups of problems, redo any problem you missed and begin Posttest A on page 127.

| Problem Numbers | Skill Area | Practice Pages |
|---|---|---|
| 1, 2, 3, 4, 5 | understanding percents | 101–106 |
| 6, 7, 8, 9, 10, 11 | finding a percent | 107–112 |
| 12, 13, 14, 15, 16 | finding what percent one number is of another | 113–118 |
| 17, 18, 19, 20, 21 | finding a number when a percent is given | 119–122 |

# Understanding Percents

**Percent** is a common term in the business world. Commissions, interest, mark-ups, discounts, credit card charges, and tax rates are all calculated with percent.

Percent is one more way—besides fractions and decimals—of describing a part of a whole. Remember that with fractions, the denominator tells how many parts a whole is divided into. Any whole number except for zero can be a denominator. With decimals a whole is divided into tenths, hundredths, thousandths, ten-thousandths, and so on. Percent is more limited. With percent the whole is always divided into 100 parts. In fact, the word *percent* means "by the hundred or per one hundred." Percent is shown with the sign %.

Think about these examples of percents.

EXAMPLES

35¢ is $\frac{35}{100}$ of a dollar or 0.35 of a dollar or 35% of a dollar

7 is $\frac{1}{2}$ of 14 or 0.5 of 14 or 50% of 14

9 is $\frac{1}{4}$ of 36 or 0.25 of 36 or 25% of 36

If a pizza is cut into eight slices and you have all eight slices, you have $\frac{8}{8}$ of the pizza or 1 whole pizza or 100% of the pizza.

**Fill in each blank.**

**1.** Percent means that a whole has been divided into ______ equal parts.

**2.** 49¢ is $\frac{49}{100}$ of a dollar or ______% of a dollar.

**3.** 75% of something means 75 of the ______ equal parts of something.

**4.** If every registered student attends a night class, you can say that ______% of the students are there.

**5.** If Gloria gets every problem right on a math quiz, you can say that she got ______% of the problems right.

**6.** If Bernard gets only half of the problems right on a math quiz, you can say that he got ______% of the problems right.

# Changing Decimals to Percents

Percent is similar to a two-place decimal. To change a decimal to a percent, move the decimal point two places to the *right* and write the percent sign (%). If the decimal point moves to the end of the number, it is not necessary to write it. You may have to add zeros.

| EXAMPLES Decimal | | Percent | |
|---|---|---|---|
| 0.35 | = 0.35 | = 35% | = 35 parts out of 100 |
| 0.8 | = 0.80 | = 80% | = 80 parts out of 100 |
| 0.04 | = 0.04 | = 4% | = 4 parts out of 100 |
| $0.12\frac{1}{2}$ | $= 0.12\frac{1}{2}$ | $= 12\frac{1}{2}\%$ | $= 12\frac{1}{2}$ parts out of 100 |
| 0.0007 | = 0.00 07 | = 0.07% | = 0.07 parts out of 100 |

**Write each decimal as a percent.**

**1.** 0.32 = 0.09 = 0.6 = 0.136 =

**2.** 0.005 = $0.37\frac{1}{2}$ = $0.08\frac{1}{3}$ = 0.045 =

**3.** 0.0016 = 0.0003 = 0.025 = $0.33\frac{1}{3}$ =

**4.** 0.125 = 0.0375 = 0.009 = 0.2 =

**5.** See how quickly you can fill in the following tables. You will save time later if you memorize these common decimals and percents.

| Decimal | Percent |
|---|---|
| 0.1 | |
| 0.3 | |
| 0.7 | |
| 0.9 | |

| Decimal | Percent |
|---|---|
| 0.2 | |
| 0.4 | |
| 0.6 | |
| 0.8 | |

| Decimal | Percent |
|---|---|
| 0.01 | |
| 0.25 | |
| 0.5 | |
| 0.75 | |

# Changing Percents to Decimals

To change a percent to a decimal, drop the percent sign (%) and move the decimal point two places to the *left*. You may have to add zeros.

EXAMPLES

| Percent | | Decimal | |
|---|---|---|---|
| 6% | = .06 | = 0.06 | = 6 parts out of 100 |
| 30% | = .30 | = 0.3 | = 3 parts out of 10 or 30 parts out of 100 |
| 150% | = 1.50 | = 1.5 | = 1 whole and 5 parts out of 10 |
| 0.9% | = .009 | = 0.009 | = 9 parts out of 1,000 |
| $37\frac{1}{2}\%$ | $= .37\frac{1}{2}$ | $= 0.37\frac{1}{2}$ | $= 37\frac{1}{2}$ parts out of 100 |

**Write each percent as a decimal.**

1. 20% =     35% =     8% =     60% =

**2.** 3.5% =     0.4% =     0.03% =     21.6% =

**3.** $62\frac{1}{2}\% =$     $6\frac{2}{3}\% =$     2.8% =     19% =

**4.** 7% =     1.5% =     200% =     14.2% =

**5.** See how quickly you can fill in the following tables. You will save time later if you memorize these common percents and decimals.

| Percent | Decimal |
|---|---|
| 50% | |
| 25% | |
| 75% | |
| 20% | |

| Percent | Decimal |
|---|---|
| 5% | |
| 1% | |
| 100% | |
| 80% | |

| Percent | Decimal |
|---|---|
| 37.5% | |
| 62.5% | |
| 87.5% | |
| 12.5% | |

# Changing Fractions to Percents

There are two ways to change a fraction to a percent as shown below.

EXAMPLE Change $\frac{3}{4}$ to a percent.

**Method 1** Multiply the fraction by 100%.

$$\frac{3}{\cancel{4}_1} \times \frac{\cancel{100\%}^{25}}{1} = \frac{75\%}{1} = \mathbf{75\%}$$

**Method 2** Divide the denominator of the fraction into the numerator and move the point two places to the right.

$$\frac{3}{4} = 4\overline{)3.00}^{\,.75} = \mathbf{75\%}$$

**Write each fraction as a percent.**

| | | | |
|---|---|---|---|
| **1.** $\frac{2}{5} =$ | $\frac{1}{4} =$ | $\frac{1}{3} =$ | $\frac{3}{8} =$ |
| **2.** $\frac{6}{25} =$ | $\frac{2}{3} =$ | $\frac{5}{6} =$ | $\frac{1}{8} =$ |
| **3.** $\frac{9}{10} =$ | $\frac{7}{8} =$ | $\frac{11}{20} =$ | $\frac{5}{12} =$ |
| **4.** $\frac{1}{6} =$ | $\frac{4}{5} =$ | $\frac{7}{10} =$ | $\frac{1}{12} =$ |
| **5.** $\frac{5}{8} =$ | $\frac{4}{9} =$ | $\frac{3}{7} =$ | $\frac{9}{20} =$ |
| **6.** $\frac{4}{25} =$ | $\frac{3}{10} =$ | $\frac{3}{5} =$ | $\frac{9}{50} =$ |

# Changing Percents to Fractions

To change a percent to a fraction, write the percent as a fraction with 100 as the denominator and reduce.

**EXAMPLE 1** Change 85% to a fraction.

**STEP 1** Write the percent as a fraction with 100 as the bottom number. $\frac{85}{100}$

**STEP 2** Reduce. $\frac{85 \div 5}{100 \div 5} = \mathbf{\frac{17}{20}}$

**EXAMPLE 2** Change $8\frac{1}{3}$% to a fraction.

**STEP 1** Write the percent as a fraction with 100 as the denominator. $\frac{8\frac{1}{3}}{100}$

**STEP 2** You can rewrite this fraction as a division problem. $8\frac{1}{3} \div 100$

**STEP 3** Change the mixed number to an improper fraction. $\frac{25}{3} \div \frac{100}{1}$

**STEP 4** Invert the divisor and multiply, canceling where possible. $\frac{\overset{1}{\cancel{25}}}{3} \times \frac{1}{\underset{4}{\cancel{100}}} = \mathbf{\frac{1}{12}}$

**Write each percent as a common fraction.**

**1.** 35% = 20% = $12\frac{1}{2}$% = 6% =

**2.** $16\frac{2}{3}$% = 1% = 90% = $37\frac{1}{2}$% =

**3.** 12% = 99% = $66\frac{2}{3}$% = $4\frac{1}{2}$% =

**4.** 80% = $33\frac{1}{3}$% = 4% = $8\frac{1}{3}$% =

# Common Fractions, Decimals, and Percents

After you have filled in the table on this page, check your answers. Then memorize the table. These are the most common fractions, decimals, and percents. You will save time later on if you know what each of them is equal to.

| Fraction | Decimal | Percent | Fraction | Decimal | Percent |
|---|---|---|---|---|---|
| $\frac{1}{2}$ = | 0.5 = | 50% | $\frac{1}{8}$ = | | |
| $\frac{1}{4}$ = | | | $\frac{3}{8}$ = | | |
| $\frac{3}{4}$ = | | | $\frac{5}{8}$ = | | |
| | | | $\frac{7}{8}$ = | | |
| $\frac{1}{5}$ = | | | $\frac{1}{10}$ = | | |
| $\frac{2}{5}$ = | | | $\frac{3}{10}$ = | | |
| $\frac{3}{5}$ = | | | $\frac{7}{10}$ = | | |
| $\frac{4}{5}$ = | | | $\frac{9}{10}$ = | | |
| $\frac{1}{3}$ = | | | $\frac{1}{6}$ = | | |
| $\frac{2}{3}$ = | | | $\frac{5}{6}$ = | | |

**Do your work here.**

# Finding a Percent of a Number

To find a percent of a number, change the percent to a decimal or to a fraction and multiply.

**EXAMPLE 1** Find 25% of 80.

Method 1

**STEP 1** Change the percent to a decimal. $25\% = .25$

**STEP 2** Multiply.

$$\begin{array}{r} 80 \\ \times\ .25 \\ \hline 400 \\ 160\phantom{0} \\ \hline 20.00 \end{array} = \mathbf{20}$$

Method 2

**STEP 1** Change the percent to a fraction. $25\% = \frac{1}{4}$

**STEP 2** Multiply.

$$\frac{1}{\cancel{4}_1} \times \frac{\cancel{80}^{20}}{1} = \frac{20}{1} = \mathbf{20}$$

**Use the method that you find easier to solve the following.**

**1.** 5% of 120 =     7% of 965 =     10% of 780 =

**2.** 20% of 36 =     15% of 50 =     40% of 60 =

**3.** 75% of 680 =     80% of 500 =     50% of 418 =

**4.** 35% of 480 =     65% of 620 =     85% of 940 =

**5.** 2.6% of 390 =     0.8% of 56 =     1.8% of 753 =

If you want to multiply by a complex percent like $16\frac{2}{3}\%$, it is easiest to change the percent to the fraction that it is equal to (from the table on page 106) and then multiply.

**EXAMPLE 2** Find $16\frac{2}{3}\%$ of 42.

**STEP 1** Change the complex percent to a fraction. $16\frac{2}{3}\% = \frac{1}{6}$

**STEP 2** Multiply. $\frac{1}{\cancel{6}_1} \times \frac{\cancel{42}^7}{1} = \frac{7}{1} = \mathbf{7}$

If you do not know the fractional value of a complex percent, first write the percent as a fraction with a denominator of 100. (See Example 2, Step 1, page 105.)

**EXAMPLE 3** Find $6\frac{2}{3}\%$ of 45.

**STEP 1** Change the complex percent to an improper fraction with a denominator of 100. $\frac{6\frac{2}{3}}{100}$

**STEP 2** Rewrite the fraction as a division problem and simplify. $6\frac{2}{3} \div 100 = \frac{20}{3} \times \frac{1}{100}$

**STEP 3** Multiply 45 by the simplified percent. $\frac{\cancel{20}^1}{\cancel{3}_1} \times \frac{1}{\cancel{100}_5} \times \frac{\cancel{45}^{15}}{1} = \frac{15}{5} = \mathbf{3}$

**Note:** In the last example, you can first simplify the complex fraction. $6\frac{2}{3}\%$ simplifies to $\frac{1}{15}$. It is easier, however, to leave the percent in the form $\frac{20}{3} \times \frac{1}{100}$.

**Solve the following.**

**6.** $33\frac{1}{3}\%$ of 75 = $\qquad$ $12\frac{1}{2}\%$ of 96 = $\qquad$ $16\frac{2}{3}\%$ of 84 =

**7.** $37\frac{1}{2}\%$ of 720 = $\qquad$ $83\frac{1}{3}\%$ of 630 = $\qquad$ $1\frac{1}{2}\%$ of 200 =

**8.** $5\frac{1}{4}\%$ of 400 = $\qquad$ $66\frac{2}{3}\%$ of 90 = $\qquad$ $87\frac{1}{2}\%$ of 200 =

**9.** $8\frac{1}{3}\%$ of 36 = $\qquad$ $6\frac{1}{4}\%$ of 60 = $\qquad$ $62\frac{1}{2}\%$ of 176 =

# Using Shortcuts

Look at the following percent list and the fraction that each percent equals.

| 50% | $33\frac{1}{3}\%$ | 25% | 20% | $16\frac{2}{3}\%$ | $12\frac{1}{2}\%$ | 10% |
|---|---|---|---|---|---|---|
| $\frac{1}{2}$ | $\frac{1}{3}$ | $\frac{1}{4}$ | $\frac{1}{5}$ | $\frac{1}{6}$ | $\frac{1}{8}$ | $\frac{1}{10}$ |

Notice that each percent in the list is equal to a fraction with a numerator of 1.

To find a percent of a number with any percent on this list, simply divide by the denominator of the fraction that the percent equals (Method 3 below).

EXAMPLE What is 25% of 120?

Method 1 $0.25 \times 120 = 30.00 = \mathbf{30}$

Method 2 $\frac{1}{\cancel{4}_1} \times \frac{\cancel{120}^{30}}{1} = \frac{30}{1} = \mathbf{30}$

Method 3 $120 \div 4 = \mathbf{30}$

The shortcut Method 3 is the same as the canceling step from Method 2.

**Use a shortcut method to solve each of the following problems in your head.**

**1.** 50% of 48 = $\qquad$ $33\frac{1}{3}\%$ of 690 = $\qquad$ 20% of 35 =

**2.** $12\frac{1}{2}\%$ of 720 = $\qquad$ 10% of 1,300 = $\qquad$ $16\frac{2}{3}\%$ of 240 =

**3.** 20% of 450 = $\qquad$ 25% of 1,600 = $\qquad$ $33\frac{1}{3}\%$ of 90 =

**4.** 10% of 170 = $\qquad$ $12\frac{1}{2}\%$ of 560 = $\qquad$ 20% of 1,500 =

**5.** $16\frac{2}{3}\%$ of 720 = $\qquad$ 50% of 482 = $\qquad$ 25% of 3,600 =

# Finding a Percent of a Number: Applying Your Skills

When you find a *percent of* a number, you find a *part of* that number.

**Solve and write the correct label, such as $ or pounds, next to each answer.**

1. During a period of 30 working days, Deborah was late 80% of the time. How many days was she late?

2. For selling his studio apartment, Martin had to pay the real estate agent a commission of 5%. The studio sold for $95,000. What was the agent's commission?

3. Linda is a customer service representative. She must answer at least 15% of the customer e-mails her company receives each day. Today, there were 40 e-mails awaiting response. How many e-mails must she answer?

4. A local electricians' union has 540 members. If 65% of the members attended the last meeting, how many members went to the meeting?

5. Doreen's night class was supposed to meet 30 times. Because of snow emergencies, 20% of the classes were canceled. How many of Doreen's classes were canceled?

6. The Smiths spend $33\frac{1}{3}\%$ of their take-home pay for food. If they take home $882 a week, how much do they spend each week for food?

7. Calvin made a down payment of $12\frac{1}{2}\%$ on a used car that cost $4,800. How much was the down payment?

8. Joe was supposed to work 250 days last year. He was absent 4% of the time because he was sick. How many days of work did he miss?

9. Ruby used to weigh 150 pounds. She went on a diet and lost 6.5% of her weight. How many pounds did she lose?

# Estimating Percent Answers

Remember that 100% is one whole. A percent *less than* 100% is less than 1. A percent less than 100% is equivalent to a proper fraction.

*Less than* 100% of a number is *part of* the number.

**EXAMPLES** 25% of 32 = 8    50% of \$18 = \$9    13.5% of 200 = 27

When you calculate *more than* 100% of a number, you get an amount that is *greater than* that number.

**EXAMPLES** 150% of 40 = 60    200% of 19 = 38    110% of 50 = 55

**In each box write the symbol $<$, $>$, or $=$ to make the statement true.**

1. 4% of \$600 $\square$ \$600    105% of 80 $\square$ 80    90% of 270 $\square$ 270
2. 210% of 70 $\square$ 70    100% of 2,390 $\square$ 2,390    1.5% of \$40 $\square$ \$40

When you find a percent of a number, sometimes you can estimate an answer by using an *easier* percent. For example, to estimate 9.6% of 400, find 10% of 400.

**EXAMPLE** Estimate an answer to 9.6% of 400.

Round 9.6% to 10%. The estimate is 10% of 400 = $0.1 \times 400$ = **40.**
The exact answer is 38.4.

Sometimes it is useful to round the number you want to find a percent of. For example, to estimate $33\frac{1}{3}\%$ of \$14.95, find $33\frac{1}{3}\%$ of \$15.

**EXAMPLE** Estimate an answer to $33\frac{1}{3}\%$ of \$14.95.

Round \$14.95 to \$15. The estimate is $33\frac{1}{3}\%$ of \$15 or $\frac{1}{3}$ of \$15 = **\$5.**
The exact answer is \$4.98.

**For each problem choose the best estimate. Then calculate both the exact answer and the estimate.**

3. 19.5% of 115 ≈    **a.** 10% of 115 =    **b.** 15% of 115 =    **c.** 20% of 115 =
4. 4.8% of 70 ≈    **a.** 50% of 70 =    **b.** 10% of 70 =    **c.** 5% of 70 =
5. 16% of 2,477 ≈    **a.** 16% of 2,000 =    **b.** 16% of 2,500 =    **c.** 16% of 3,000 =
6. 74.2% of 360 ≈    **a.** 75% of 360 =    **b.** 740% of 360 =    **c.** 50% of 360 =
7. 125% of 816 ≈    **a.** 125% of 1,000 =    **b.** 125% of 850 =    **c.** 125% of 800 =
8. 38% of \$65 ≈    **a.** 30% of \$65 =    **b.** 40% of \$65 =    **c.** 50% of \$65 =

# Solving Two-Step Problems

Many applications of finding a percent of a number require two steps. First, find the percent of a number. Then add or subtract. For example, after you calculate sales tax on an item, *add* the tax to the original price. When you calculate the amount you can save on a sale item, *subtract* the savings from the original price.

**Read each problem carefully to decide whether to add or subtract. Then use the suggestion that follows each problem to estimate the answer.**

1. A jacket selling for $48 was on sale at 20% off. Find the sale price of the jacket. To estimate, round the original price to the nearest $10.

2. Paul earns $380 each week at his part-time job. His employer withholds 18% of Paul's pay for taxes and social security. How much does Paul take home each week? To estimate, round the percent to the nearest ten percent.

3. Elizabeth takes home $576 each week at her job as a clerk-typist. If she gets an 8% raise, how much will she take home each week? To estimate, round her pay to the nearest $10, and round the percent to the nearest 10%.

4. The Chan family bought their house for $118,000. They sold it 15 years later for 22% more than they paid. For how much did they sell their house? To estimate, round the price to the nearest ten thousand and the percent to the nearest 10%.

5. According to the last census, the population of Franklin County was 253,000. By the time the next census is taken, the population is expected to decrease by 4.5%. What will be the new population of the county? To estimate, round the population to the nearest ten thousand and the percent to the nearest percent.

6. A computer that sold for $695 last year is now on sale for 12% less. What is the price of the computer this year? To estimate, round the price to the nearest hundred and the percent to the nearest 10%.

7. A pair of winter boots originally selling for $79.90 was on sale for 15% off. Find the sale price of the boots. To estimate, round the original price to the nearest $10.

8. Of the 320 employees in the factory where Manuel works, 78% participate in the company savings plan. How many employees do *not* participate in the plan? To estimate, round the percent to the nearest 10%.

9. The day Gus' Coffee Shop opened, there were 82 diners for lunch. The next day, the number of lunchtime diners increased by 66%. How many people ate lunch at Gus' on the second day? Express the answer to the nearest whole number. To estimate, round the number of diners to the nearest ten and the percent to the nearest 10%.

# Finding What Percent One Number Is of Another

To find what percent one number is of another, make a fraction by putting the part (usually the smaller number) over the whole. Reduce the fraction and change it to a percent.

**EXAMPLE** 9 is what percent of 45?

**STEP 1** Put the part over the whole and reduce. $\frac{9}{45} = \frac{1}{5}$

**STEP 2** Divide the denominator into the numerator (see page 104). $5\overline{)1.00}$ = 0.20

**STEP 3** Change the decimal answer to a percent by moving the decimal point two places to the right and adding a percent (%) sign. 0.20 = **20%**

**Solve the following.**

1. 8 is what percent of 16? 15 is what percent of 60? 27 is what percent of 81?

2. 9 is what percent of 90? 12 is what percent of 72? 40 is what percent of 320?

3. 16 is what percent of 20? 14 is what percent of 35? 32 is what percent of 48?

4. 75 is what percent of 90? 33 is what percent of 44? 56 is what percent of 64?

5. 35 is what percent of 40? 45 is what percent of 50? 160 is what percent of 200?

6. 14 is what percent of 200? | 24 is what percent of 400? | 45 is what percent of 500?

7. 7 is what percent of 20? | 19 is what percent of 50? | 8 is what percent of 25?

8. 21 is what percent of 36? | 19 is what percent of 19? | 12 is what percent of 27?

9. 60 is what percent of 72? | 24 is what percent of 56? | 126 is what percent of 140?

10. 4 is what percent of 200? | 18 is what percent of 300? | 12 is what percent of 144?

11. 15 is what percent of 75? | 84 is what percent of 105? | 27 is what percent of 120?

12. 12 is what percent of 72? | 2,600 is what percent of 10,000? | 792 is what percent of 200,000?

# Finding What Percent One Number Is of Another: Applying Your Skills

To find what percent one number is of another, first make a fraction with the *part* over the *whole*. Then reduce and change the fraction to a percent.

**Solve each problem.**

1. There are 24 members in a community block association. Only 18 of them came to their winter meeting. What percent of the members attended the meeting?

2. Joaquin makes $900 a week. A raise of $36 a week represents what percent of his salary?

3. Floria wants to buy a dress that costs $96. She has saved $60 for the dress. What percent of the price of the dress has she saved?

4. Matthew got 36 problems right out of a total of 40 problems on a test. What percent of the problems did he get right?

5. On a loan of $3,500, Manny had to pay $280 interest. The interest was what percent of Manny's loan?

6. Carl weighed 180 pounds. After two months of dieting and exercising, he lost 9 pounds. What percent of his weight did he lose?

7. As a paralegal, Margaret has to type a document that is 96 pages long. If she has already typed 32 pages, what percent of the document has she typed?

8. In the last problem, what percent of the document does Margaret have left to type? (**Hint:** Think of the whole document as 100%.)

9. The Johnson family spends $1,040 a month on rent. If their monthly income is $4,160, what percent of their income goes to rent?

10. In an office with 20 employees, there are 11 men. What percent of the office workers are men?

11. In the last problem, what percent of the office workers are women?

12. Carlos owes his sister $2,000. If he has paid back $1,250, what percent of the amount that he owes has he paid back?

13. The price of a dozen large eggs went up $0.08 from $1.28. The increase represents what percent of the original price?

14. The Rivera family makes $720 a week. If they spend $180 each week for food, what percent of their income goes for food?

15. For the Rivera family in the last problem, what percent of their weekly income is left for all expenses except food?

16. George bought a used car last year for $3,520. This year it is worth $440 less. The decrease in value represents what percent of the price George paid?

17. Last year the population of Middleville was 16,000 people. In one year the population has increased by 800 people. The increase represents what percent of last year's population?

18. Rachel belongs to a car pool. She has driven her car to work 8 days out of the past 18. What percent of that time did she drive her car to work?

19. An inch represents what percent of a foot?

20. A foot is what percent of a yard?

# Finding a Percent of Change

A common application of percent is to find a percent of change. First, find the amount of the change. Next, make a fraction with the change over the original (earlier) amount. Finally, change that fraction to a percent.

EXAMPLE 1 **The price of a jacket dropped from $30 to $24. By what percent did the price drop?**

**STEP 1** Find the amount of the change in price. $\$30 - \$24 = \$6$

**STEP 2** Make a fraction with the change over the original price and reduce. $\frac{\$6}{\$30} = \frac{1}{5}$

**STEP 3** Change $\frac{1}{5}$ to a percent. $\frac{1}{5} = \mathbf{20\%}$

The *original amount* may be larger or smaller than the other amount in a problem. In the last example, the original price, $30, was larger than the sale price, $24. Think about the next example.

EXAMPLE 2 **The number of students in Maria's evening Spanish class went from 16 when the class began to 20 one month later. By what percent did the number of students change?**

**STEP 1** Find the amount of the change. $20 - 16 = 4$

**STEP 2** Make a fraction with the change over the original. $\frac{4}{16} = \frac{1}{4}$

**STEP 3** Change $\frac{1}{4}$ to a percent. $\frac{1}{4} = 25\%$

**Solve each problem. Remember to write the amount of change over the original amount.**

**1.** A high definition TV that originally sold for $380 was on sale for $285.

**a.** By how much did the price of the TV change?

**b.** What was the original price of the TV?

**c.** Make a fraction with the change in price over the original price. Then reduce.

**d.** By what percent did the price of the TV drop?

2. When Miguel started working at Acme Plastics, there were 80 full-time employees. Two years later there were 104 full-time employees at the factory.
   **a.** By how many did the number of full-time employees increase?
   **b.** What was the original number of employees when Miguel started working at Acme?
   **c.** Make a fraction with the increase over the original number of employees.
   **d.** By what percent did the number of full-time employees increase while Miguel worked at Acme?

3. Anna's weekly salary was $800. To spend more time with her children, she has decided to work part-time. Now her weekly salary will be $632. By what percent did her salary decrease?

4. At the beginning of the football season, 1,600 people attended a high school game. After the team lost several games, the attendance was down to 1,120 people. By what percent did the attendance drop?

5. Last year a town budget was $2.4 million. This year the budget will be $2.7 million. By what percent did the budget increase from last year?

6. Mr. Park pays a distributor $210 for a home entertainment system. He charges his customers $280 for the system. The markup (the additional amount Mr. Park charges) is what percent of the price Mr. Park pays the distributor?

7. The Lincoln family bought their house for $64,000. They sold the house twenty years later for $144,000. By what percent did the value of the house increase?

8. Mr. Philips bought his used truck for $8,100. Now he wants to sell his truck, but he can get only $4,500 for it. By what percent has the value of his truck dropped?

9. One winter the price of heating fuel increased from $2.80 a gallon to $3.15. By what percent did the price change that season?

10. Under pressure from his wife, his boss, and his doctor, Tim decided to lose weight. In January he weighed 212 pounds. After six months of exercising and dieting, his weight was 180 pounds. What percent of Tim's January weight did he lose? Express the answer to the nearest percent.

# Finding a Number When a Percent of It Is Given

If a percent of a number is given and you are looking for the whole number, change the percent into either a fraction or a decimal and divide it into the number you have.

EXAMPLE 20% of what number is 16?

Method 1

**STEP 1** Change the percent to a fraction. $20\% = \frac{1}{5}$

**STEP 2** Divide the number you have by the fraction. $16 \div \frac{1}{5} = \frac{16}{1} \times \frac{5}{1} = \frac{80}{1} = \mathbf{80}$

Method 2

**STEP 1** Change the percent to a decimal. $20\% = 0.2$

**STEP 2** Divide the number you have by the decimal. $0.2\overline{)16.0}$ = **80.**

**Use the method you find easier to solve the following.**

1. 25% of what number is 8? 50% of what number is 45?

2. 75% of what number is 48? 60% of what number is 75?

3. 40% of what number is 60? 15% of what number is 12?

4. 10% of what number is 6.3? 35% of what number is 8.4?

5. $33\frac{1}{3}\%$ of what number is 25?

$12\frac{1}{2}\%$ of what number is 160?

6. $16\frac{2}{3}\%$ of what number is 300?

8.5% of what number is 119?

7. $37\frac{1}{2}\%$ of what number is 24?

$83\frac{1}{3}\%$ of what number is 150?

8. $66\frac{2}{3}\%$ of what number is 320?

$87\frac{1}{2}\%$ of what number is 210?

9. $62\frac{1}{2}\%$ of what number is 120?

12.6% of what number is 189?

10. 80% of what number is 244?

$33\frac{1}{3}\%$ of what number is 350?

11. $83\frac{1}{3}\%$ of what number is 365?

20% of what number is 175?

# Finding a Number When a Percent of It Is Given: Applying Your Skills

In these word problems, you have a *part* and a percent rate. You are looking for the *whole*. Change the percent to either a decimal or a fraction. Then divide into the part you have. If the percent rate in the problem is less than 100%, your answer (the whole) will be larger than the part.

**Solve and write the correct label, such as $ or pounds, next to each answer.**

1. Morris has saved $95 toward the purchase of a new computer printer. The $95 is 50% of the price. What is the price of the printer?

2. Lois got 27 problems right on a math test. If this is 90% of the test, how many problems were on the test?

3. Pat made a down payment of $1,080 on a used car. The $1,080 is 15% of the price of the car. Find the price of the car.

4. There were 130 people in a school auditorium to listen to a political candidate. This represents $33\frac{1}{3}$% of the capacity of the auditorium. How many seats are in the auditorium?

5. Phil had to pay $4.80 in tax for a new pair of pants. If the sales tax in Phil's state is 8%, what was the price of the pants?

6. There are 78 members at a club meeting. If they represented 65% of the total membership, how many members does the club have?

7. Mavis made $620 in commissions one month for selling shoes. Her commission rate is 5% of the total amount that she sells. What was the value of the shoes she sold that month?

8. When the Greens bought their house, they made a down payment of $25,500, which was 15% of the total price of the house. Find the price of the house.

9. Sonia works in quality control at a factory that makes car radios. One week she found 18 defective radios. This represents 0.8% of the number of radios produced. How many radios did the factory produce that week?

10. Last week 3,000 people came to hear a presidential candidate speak in Central City. If they represent $16\frac{2}{3}\%$ of the people who live in Central City, how many people live there?

11. David has driven 2,550 miles on his way from New York to San Francisco. If he has driven 85% of the total distance, what is the distance from New York to San Francisco?

12. Tom had to pay $240 in interest on a loan. The interest was 12% of the total loan. How much was the loan?

13. Mr. Moore pays $21 to a supplier for the hand drills he sells in his hardware store. $21 is 60% of the price he charges his customers. How much does Mr. Moore charge for a hand drill at his store?

14. Walter's weekly take-home pay is $709.80, which is 78% of his gross pay (before deductions). What is Walter's weekly gross pay?

15. In the last problem, what are the weekly deductions from Walter's gross pay?

# Recognizing Types of Percent Problems

In the statement, "6 is 20% of 30," the number 6 is the *part*, the number 30 is the *whole*, and 20% is the *percent*. Three problems can be made from these numbers.

**EXAMPLE 1** *Finding the part:* What is 20% of 30?

This is the most common type of percent problem.

To solve Example 1, change the percent to a fraction or a decimal and *multiply* by the *whole*.

$20\% = 0.2$
$0.2 \times 30 = \mathbf{6}$

**EXAMPLE 2** *Finding the percent:* 6 is what percent of 30?

This is the easiest type of percent problem to recognize. The percent is not given.

To solve Example 2, make a fraction with the *part* over the *whole* and change the fraction to a percent.

$\frac{6}{30} = \frac{1}{5} = \mathbf{20\%}$

**EXAMPLE 3** *Finding the whole:* 20% of what number is 6?

This is the least common type of percent problem.

To solve Example 3, change the percent to a decimal or a fraction and *divide* into the *part*.

$20\% = 0.2$

$0.2\overline{)6.0}$ = **30.**

**Next to each problem, write *P* for finding the part, % for finding the percent, or *W* for finding the whole. Then solve each problem.**

1. 18 is what percent of 36? What is 30% of 110?

2. Find $12\frac{1}{2}\%$ of 720. What percent of 50 is 10?

3. 15 is 25% of what number? What is 85% of 400?

4. What is $37\frac{1}{2}\%$ of 40? 48 is what percent of 60?

5. What percent of 60 is 20? 230 is 50% of what number?

6. Of the 500 employees at Ajax Electronics, only 6% go to work by public transportation. How many employees at Ajax use public transportation to get to work?

7. The Moore family expects to spend $1,200 on their summer vacation. So far they have saved $900 toward their vacation. What percent of the cost have they saved?

8. The sales tax rate in Linda's state is 5%. How much tax does she have to pay on a shirt that costs $29?

9. Juan saves 10% of his take-home pay. He puts $320 in his savings account each month. What is his monthly take-home pay?

10. The total bill for Paul and Dorothy's dinner was $48. They left a tip of $9.60. The tip was what percent of the total bill?

11. Of the 360 people who took a qualifying exam for an inspector's job, 55% of the people passed. How many people passed?

12. A school sold 1,020 tickets for a benefit concert. The ticket sales were 85% of the seating capacity of the school auditorium. How many seats are there in the auditorium?

# Cumulative Review

This review covers the material you have studied so far in this book. When you finish, check your answers at the back of the book.

**1.** $4\frac{5}{8} + 9\frac{1}{4} =$

**2.** $12 - 7\frac{3}{5} =$

**3.** $5\frac{1}{4} \times \frac{1}{3} =$

**4.** Round 1.297 to the nearest hundredth.

**5.** $0.25 \times 1.3 =$

**6.** Find $4.6 \div 3$ to the nearest tenth.

**7.** Write 0.123 as a percent.

**8.** Write 5.6% as a decimal.

**9.** Write $\frac{7}{12}$ as a percent.

**10.** Write 84% as a fraction.

**11.** Write $31\frac{1}{4}\%$ as a fraction.

**12.** Find 9% of 270.

**13.** Find 11.3% of 460.

**14.** Which of the following is the same as $12\frac{1}{2}\%$ of 72?

**a.** $72 \div 4$

**b.** $72 \div 6$

**c.** $72 \div 8$

**d.** $72 \div 12$

**15.** Bill works at an electronics store. He is allowed to give up to 15% in discounts to customers. What is the lowest price that he can offer on a digital camera that normally sells for $165?

**16.** There are 24,000 registered voters in Millville. If only $37\frac{1}{2}\%$ of them voted in the last election, how many of the registered voters did *not* vote?

**17.** 45 is what percent of 72?

**18.** 17 is what percent of 200?

**19.** The Millers want to buy a mobile home that costs $36,000. They have already saved $27,000. What percent of the total cost have they saved?

**20.** Of the 32 employees in an office, 20 of them are women. What percent of the employees are women?

**21.** 40% of what number is 80?

**22.** $12\frac{1}{2}$% of what number is 48?

**23.** Alfredo has driven 273 miles, which is 65% of the total distance he has to drive. Find the total distance he has to drive.

**24.** By Thursday, the cutters and sewers in a clothing factory had made 280 jackets. This is $87\frac{1}{2}$% of the total number of jackets they were expected to make for the week. What is the total number of jackets the workers had to make that week?

## CUMULATIVE REVIEW CHART

If you missed more than one problem on any group below, review the practice pages for those problems. Then redo the problems you got wrong before going on to Posttest A. If you had a passing score, redo any problem you missed and begin Posttest A on page 127.

| Problem Numbers | Skill Area | Practice Pages |
|---|---|---|
| 1, 2, 3 | understanding fractions | 11–59 |
| 4, 5, 6 | understanding decimals | 64–96 |
| 7, 8, 9, 10, 11 | understanding percents | 101–106 |
| 12, 13, 14, 15, 16 | finding a percent | 107–112 |
| 17, 18, 19, 20 | finding what percent one number is of another | 113–118 |
| 21, 22, 23, 24 | finding a number when a percent is given | 119–122 |

# Posttest A

This posttest gives you a chance to check your understanding of fractions, decimals, and percents. Take your time and work each problem carefully. When you finish, check your answers and review any topics on which you need more work.

**1.** Reduce $\frac{16}{36}$ to lowest terms.

**2.** $5\frac{7}{8} + 4\frac{2}{3} + 9\frac{1}{2} =$

**3.** Is the sum of $\frac{5}{16} + \frac{7}{15}$ greater than 1, less than 1, or equal to 1?

**4.** A table that is $46\frac{1}{2}$ inches long can be made longer with an insert that is $12\frac{2}{3}$ inches long. How long is the table with the insert?

**5.** $23\frac{5}{12} + 18\frac{3}{4} =$

**6.** A copper pipe that is $75\frac{1}{4}$ inches long is how much longer than a pipe that is $68\frac{5}{16}$ inches long?

**7.** $\frac{5}{6} \times \frac{8}{15} \times \frac{3}{20} =$

**8.** $4\frac{1}{2} \times 2\frac{2}{5} \times 3\frac{3}{4} =$

**9.** Which of the following is the best estimate for the value of $4\frac{1}{3} \times 5\frac{7}{8}$?

**a.** $5 \times 6 = 30$ **b.** $4 \times 6 = 24$ **c.** $4 \times 5 = 20$

**10.** A plumber charges $29 for an hour of work. How much would he charge for a job that took him $1\frac{3}{4}$ hours to finish?

**11.** $\frac{3}{5} \div 15 =$

**12.** $4\frac{2}{3} \div 3\frac{3}{4} =$

13. Simon has to fill $\frac{3}{4}$-pound bags with sugar from a bin that contains 36 pounds of sugar. Which of the following is a reasonable estimate of the number of bags he can fill?

   **a.** exactly 36 bags **b.** fewer than 36 bags **c.** more than 36 bags

14. Henry wants to buy a new table saw for his carpentry shop. He has saved $360, which is $\frac{2}{3}$ of the price of the saw. How much does the table saw cost?

15. Write fifty and two hundred eight thousandths as a mixed decimal.

16. Write the following decimals in order from smallest to largest: 0.021, 0.12, 0.2, 0.02

17. $293.08 + 14 + 2.719 =$

18. Mary weighed 119.3 pounds in September. By January 1, she weighed 8.8 pounds more. What was her weight on January 1?

19. $40 - 0.387 =$

20. In the problem $19.28 - 1.546$, first round each number to the nearest *tenth.* Then subtract.

21. The Rodriguez family needs to drive 250 miles to visit their cousins. If they have already driven 113.8 miles, how much farther do they need to drive?

22. $2.048 \times 1.9 =$

23. In the problem $67.4 \times 0.08$, first round 67.4 to the nearest *ten.* Then multiply.

**24.** Find the cost of 4.8 kilograms of potatoes at the price of $0.75 a kilogram.

**25.** $1.274 \div 4.9 =$

**26.** $0.3474 \div 0.018 =$

**27.** Find the answer to $24 \div 1.7$ to the nearest tenth.

**28.** There are 1.6 kilometers in a mile. Find the distance in miles between two towns that are 24 kilometers apart.

**29.** Write 7.9% as a decimal.

**30.** Write $\frac{9}{16}$ as a percent.

**31.** Write $16\frac{2}{3}\%$ as a common fraction.

**32.** Find 18% of 304.

**33.** Find $8\frac{1}{3}\%$ of 144.

**34.** Which of the following is the same as 25% of 84?

**a.** $84 \div 5$
**b.** $84 \div 4$
**c.** $84 \div 3$
**d.** $84 \div 2$

**35.** Mike gets an 8% commission on the sporting goods that he sells. One day he sold sporting goods worth $2,768. Find his commission for that day.

**36.** According to his calculation, Bill has already paid 78.5% of the $11,964 credit-card debt he had at the beginning of the year. Which of the following problems is a good way to estimate the amount of his debt that he has paid off?

**a.** 80% of $12,000
**b.** 70% of $11,964
**c.** 21% of $11,000

**37.** 75 is what percent of 90?

**38.** Sandy wants to save $1,200 to get her car repaired. She has already saved $750. What percent of the money that she needs has she saved?

**39.** 35% of what number is 56?

**40.** A community has raised $200,000 for its summer youth program. This amount is 80% of the total they need. What is the total cost of the summer youth program?

# POSTTEST A CHART

Circle the number of any problem that you miss. If you missed one or less in each group below, go on to Using Number Power. If you missed more than one problem in any group, review the pages in this book.

| PROBLEM NUMBERS | SKILL AREA | PRACTICE PAGES |
| --- | --- | --- |
| 1, 2, 3, 4, 5, 6, 7, 8, 9, 10, 11, 12, 13, 14 | using fractions | 11–59 |
| 15, 16, 17, 18, 19, 20, 21, 22, 23, 24, 25, 26, 27, 28 | using decimals | 64–96 |
| 29, 30, 31, 32, 33, 34, 35, 36, 37, 38, 39, 40 | using percents | 101–124 |

# USING NUMBER POWER

# Reading a Ruler

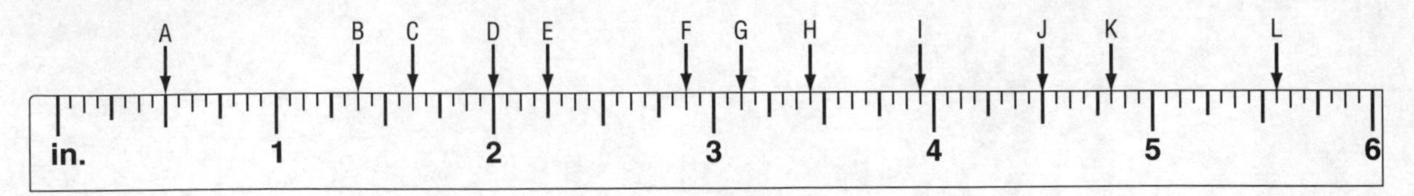


The six-inch ruler shown on this page is divided into inches (at the numbers), half inches (every eight lines), quarter inches (every four lines), eighth inches (every other line), and sixteenth inches (from one line to the next).

Remember that rulers, like books, are read from left to right. The arrow at A is $\frac{1}{2}$ inch from the left end of the ruler. The arrow at B is $1\frac{3}{8}$ inches from the left end.

**Use the ruler above to answer the following questions.**

**1.** Tell in inches how far from the left end each of the following points is.

| | |
|---|---|
| C | H |
| D | I |
| E | J |
| F | K |
| G | L |

**2.** How much farther is point D from the left end of the ruler than point C?

**3.** How much farther is point H from the left end of the ruler than point F?

**4.** How much farther is point L from the left end of the ruler than point J?

**5.** What is the distance between point A and point I?

**6.** What is the distance between point F and point L?

# Reading a Metric Ruler

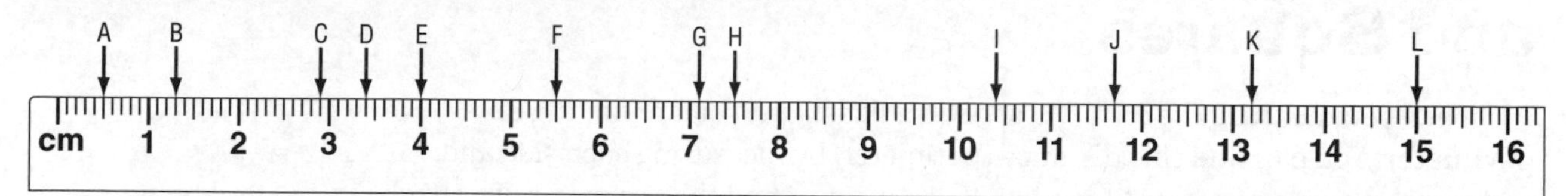


The sixteen-centimeter ruler shown on this page is divided into centimeters (at the numbers) and millimeters at each line. There are ten millimeters in a centimeter. Every fifth millimeter is a longer mark on the ruler.

The arrow at A is 5 millimeters or 0.5 centimeters from the left end of the ruler. The arrow at B is 1.3 centimeters from the left end.

**Use the ruler above to answer the following questions.**

**1.** Tell in centimeters how far from the left end each of the following points is.

| | |
|---|---|
| C | H |
| D | I |
| E | J |
| F | K |
| G | L |

**2.** How much farther is point E from the left end of the ruler than point C?

**3.** How much farther is point G from the left end of the ruler than point F?

**4.** How much farther is point K from the left end of the ruler than point J?

**5.** What is the distance between point D and point K?

**6.** What is the distance between point H and point J?

# Finding the Perimeter of Rectangles and Squares

When you measure the distance around a flat shape, you are finding its **perimeter.** You can find this distance (perimeter) by measuring each side and adding all the measurements together. If you want to find the perimeter of a shape with two pairs of equal sides (a **rectangle**), you can double the length (the long sides), double the width (the short sides), and add these two answers. This rule can be written in a short form called a formula: $P = 2l + 2w$. $P$ stands for perimeter, $l$ stands for length, and $w$ stands for width.

EXAMPLE How much tape do you need to go around the edges of a rectangle with a length of $4\frac{3}{4}$ inches and a width of $2\frac{1}{2}$ inches?

STEP 1 $2 \times 4\frac{3}{4} = \frac{\overset{1}{\cancel{2}}}{1} \times \frac{19}{\underset{2}{\cancel{4}}} = \frac{19}{2} = 9\frac{1}{2}$ inches

STEP 2 $2 \times 2\frac{1}{2} = \frac{\overset{1}{\cancel{2}}}{1} \times \frac{5}{\underset{1}{\cancel{2}}} = 5$ inches

STEP 3 $9\frac{1}{2}$ inches
$+ 5$ inches
**$14\frac{1}{2}$ inches**

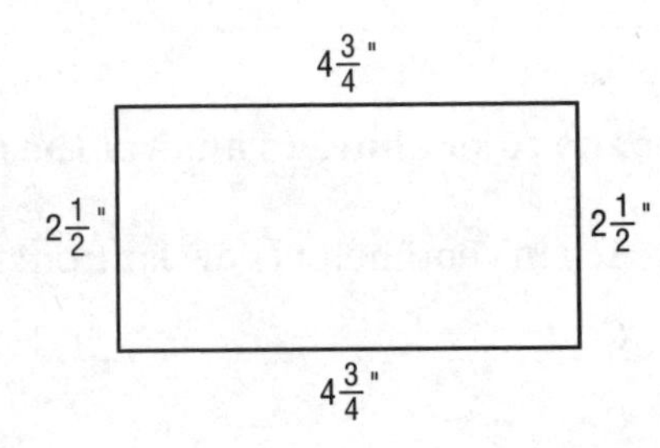


**Solve each problem. Be sure to label each answer.**

1. How much tape do you need to go around the edges of a rectangle $7\frac{2}{3}$ inches long and 3 inches wide?

2. How much weather stripping will you need to go around a window that is $6\frac{7}{8}$ feet long and $2\frac{1}{4}$ feet wide?

3. How many inches of picture frame molding are needed to go around a picture that is 11 inches long and $8\frac{1}{2}$ inches wide?

4. Carlin has a business boarding pets. The rectangular dog run in her yard is $5\frac{1}{2}$ feet wide and 36 feet long. What is the perimeter of the dog run?

5. Mr. King's garden is 6.2 meters long and 5.8 meters wide. How many meters of fencing are needed to enclose the garden?

6. If fencing costs $9.50 per meter, how much would the fencing for the garden in problem 5 cost?

7. Find the perimeter of a tabletop that is $62\frac{1}{2}$ inches long and $30\frac{1}{4}$ inches wide.

8. What is the perimeter of a photograph that is 10.3 centimeters long and 4.6 centimeters wide?

**Note:** A square is a shape with four equal sides. To find the perimeter of a square, multiply the measurement of one side by 4. The formula is $P = 4s$. $P$ stands for perimeter, and $s$ stands for one side.

9. How much fencing would be needed to enclose a square garden that measures $6\frac{1}{4}$ yards on every side?

10. Find the perimeter of a square snapshot that measures $2\frac{1}{2}$ inches on each side.

11. What is the perimeter of a square concrete patio that measures 4.25 meters on every side?

12. What is the perimeter of a square mirror that measures 0.65 meter on each side?

# Finding the Area of Rectangles and Squares

When you buy carpeting, you have to know how much space you want to cover. To figure out the answer to problems like this, you need to find **area.** Area is the amount of surface on a flat figure like a floor. Area is measured in square inches, square feet, square yards, and so on. To find the area of a rectangle, multiply the length by the width. The formula is $A = lw$. $A$ stands for the area, $l$ stands for the length, and $w$ stands for the width. Notice that $l$ and $w$ written next to each other means that they should be multiplied together.

The area of the rectangle pictured here is $5 \times 3 = 15$ square inches.

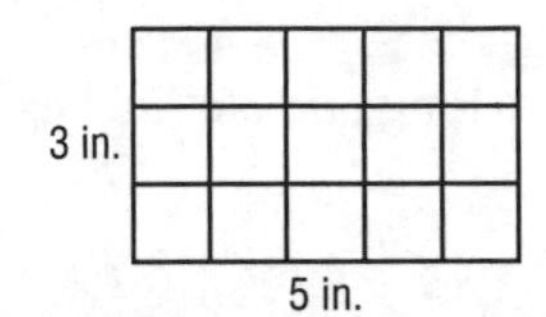


EXAMPLE How much shelf paper do you need to cover a shelf that is $4\frac{1}{2}$ feet long and $1\frac{1}{4}$ feet wide?

$4\frac{1}{2} \times 1\frac{1}{4} = \frac{9}{2} \times \frac{5}{4} = \frac{45}{8} = $ **$5\frac{5}{8}$ square feet**

**Solve each problem. Be sure to label each answer in square units.**

1. How much glass do you need to cover a coffee table that is $5\frac{1}{3}$ feet long and $3\frac{3}{4}$ feet wide?

2. How much space is covered by a tile that is 6.2 centimeters long and 2.8 centimeters wide?

3. How many square inches of plastic are needed to cover a photograph that is $7\frac{1}{2}$ inches long and 4 inches wide?

4. How many square feet are in a flower garden that is 14 feet long and 2 feet wide?

5. Find the area of a floor that is 7.2 meters long and 5.5 meters wide.

6. If one square meter of carpet costs $16.40, how much would the carpet cost to cover the floor in problem 5?

7. How many square yards of carpeting are needed to cover the floor of an entrance hall that is $3\frac{1}{3}$ yards long and $2\frac{1}{4}$ yards wide?

8. If one square yard of carpet costs $24.90, what is the cost of the carpet for the hall in problem 7?

**Note:** In a square, the length and the width are equal. To find the area of a square, multiply the measurement of one side by itself.

9. How much space can you cover with a square piece of linoleum that measures 3.6 meters on each side?

10. Find the area of a square mirror that measures 0.4 meter on each side.

11. To the nearest square foot, what is the area of a square tabletop that measures $4\frac{1}{4}$ feet on each side?

12. How much glass is needed to cover the front of a photograph that measures $6\frac{1}{2}$ inches on each side?

# Finding the Volume of a Rectangular Solid

To find out how much space is *inside* objects like boxes, suitcases, and rooms, you have to find volume. **Volume** is measured in cubic inches, cubic feet, cubic yards, and so on. To find the volume of an object like a rectangular box, multiply the length by the width by the height of the object. The formula is $V = lwh$. $V$ stands for the volume, $l$ stands for the length, $w$ stands for the width, and $h$ stands for the height.

EXAMPLE Find the volume of a cardboard box that is 8 inches long, $4\frac{1}{2}$ inches wide, and $3\frac{1}{3}$ inches high.

$$8 \times 4\frac{1}{2} \times 3\frac{1}{3} = \frac{\overset{4}{\cancel{8}}}{1} \times \frac{\overset{3}{\cancel{9}}}{\underset{1}{\cancel{2}}} \times \frac{10}{\underset{1}{\cancel{3}}} = \textbf{120 cubic inches}$$

**Solve each problem. Be sure to label each answer in cubic units.**

1. Find the volume of a box that is 3 feet long, $2\frac{1}{2}$ feet wide, and $1\frac{1}{3}$ feet high.

2. Find the volume of a water tank that is 10 feet long, 8.5 feet wide, and 6.25 feet deep.

3. How many cubic feet of air are there in a room that is 12 feet long, 9 feet wide, and $8\frac{1}{2}$ feet high?

4. Find the volume of a briefcase that is 18 inches long, 12 inches wide, and $2\frac{3}{4}$ inches deep.

5. To the nearest cubic inch, what is the volume of a laptop computer that is 14 inches long, 10.5 inches wide, and 1.5 inches high?

6. How much space is there inside a packing crate that is 16 inches long, $12\frac{1}{2}$ inches wide, and $7\frac{1}{2}$ inches high?

7. What is the volume of a matchbox that is 4 inches long, $2\frac{1}{2}$ inches wide, and $\frac{3}{4}$ inch high?

8. How many of the matchboxes in problem 7 fit inside the packing crate in problem 6?

9. What is the volume of a hole for a building foundation that is 20.4 meters long, 16.5 meters wide, and 3.8 meters deep?

10. How much concrete is in a sidewalk that is 63 feet long, 4 feet wide, and $\frac{1}{3}$ foot deep?

11. Marco is the contractor building the sidewalk in the last problem. He needs to know the volume of the concrete in cubic yards. One cubic yard is equal to 27 cubic feet. Which of the following is the best estimate of the volume of the concrete?

    **a.** a little more than 12 cubic yards
    **b.** a little more than 9 cubic yards
    **c.** about 3 cubic yards
    **d.** about 1 cubic yard

12. A reflecting pool is 30 feet long, 12 feet wide, and 0.75 foot deep. One cubic foot of space holds 7.48 gallons. When the reflecting pool is full, approximately how many gallons of water will it hold?

    **a.** 300
    **b.** 1,000
    **c.** 1,500
    **d.** 2,000

# Finding the Circumference of a Circle

The distance across a circle through its center is the **diameter.** The distance around a circle is called the **circumference.**

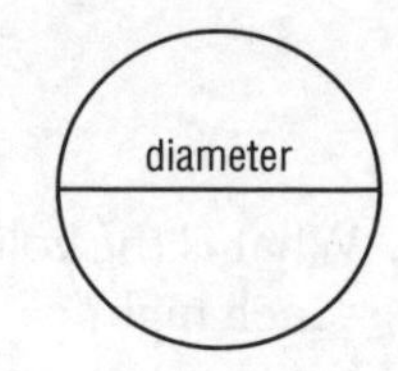


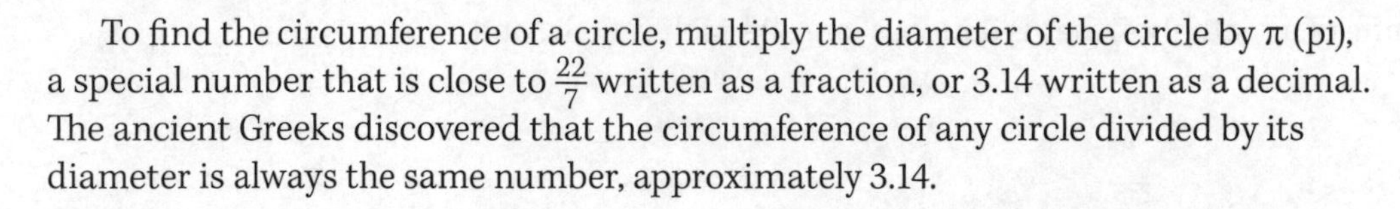

To find the circumference of a circle, multiply the diameter of the circle by $\pi$ (pi), a special number that is close to $\frac{22}{7}$ written as a fraction, or 3.14 written as a decimal. The ancient Greeks discovered that the circumference of any circle divided by its diameter is always the same number, approximately 3.14.

The formula for the distance around a circle is $C = \pi d$. $C$ is the circumference, $\frac{22}{7}$ and 3.14 are approximate values for $\pi$, and $d$ is the diameter. Notice that $\pi$ and $d$ written next to each other means that they should be multiplied together.

EXAMPLE **Find the circumference of a circular pane of glass with a diameter of 21 inches.**

$$\frac{22}{7} \times 21 = \frac{22}{\cancel{7}_1} \times \frac{\cancel{21}^3}{1} = \textbf{66 inches}$$

**Solve each problem.**

**1.** Find the circumference of a circle with a diameter of 56 yards. Use $\pi = \frac{22}{7}$.

**2.** Mr. Casa wants to put a fence around a circular space in his backyard for his small children to play in. If the space has a diameter of 35 feet, how many feet of fencing does he need? Use $\pi = \frac{22}{7}$.

**3.** If the fencing for problem 2 costs $2.15 a foot, how much will the fencing cost to enclose the space in Mr. Casa's backyard?

**4.** A circular pool has a diameter of 20 feet. Calculate the distance around the pool. Use $\pi = 3.14$.

5. What is the circumference of a hubcap with a diameter of 10 inches? Use $\pi = 3.14$.

6. Charles built a round table with a diameter of $3\frac{1}{2}$ feet. He wants to put metal stripping around the edge of the table. How many feet of stripping must he buy? Use $\pi = \frac{22}{7}$.

7. If the stripping for problem 5 costs $1.18 a foot, how much will Charles pay for the stripping to go around his table?

8. How many inches of wood are needed to make a frame for a circular mirror with a diameter of 28 inches? Use $\pi = \frac{22}{7}$.

9. Margie is an interior decorator. How much fringe does she need to trim a lampshade that has a diameter of 14 inches? Use $\pi = \frac{22}{7}$.

10. How many feet of fencing are needed to go around a circular pool that has a diameter of 30 feet? Use $\pi = 3.14$.

11. What is the perimeter of a circular picture with a diameter of 1 foot? Use $\pi = 3.14$.

12. If the framing for the picture in the last problem costs $15 a foot, find, to the nearest dollar, the cost of the framing.

# Finding the Area of a Circle

You already learned that the distance across a circle is called the diameter. The distance halfway across a circle is called the **radius**. The radius is $\frac{1}{2}$ the diameter. To find the area of a circle, multiply the radius by itself and then multiply this answer by $\pi$. The formula for finding the area of a circle is $A = \pi r^2$. $A$ is the area, $\frac{22}{7}$ or 3.14 can be used for $\pi$, and $r^2$ is the radius multiplied by itself.

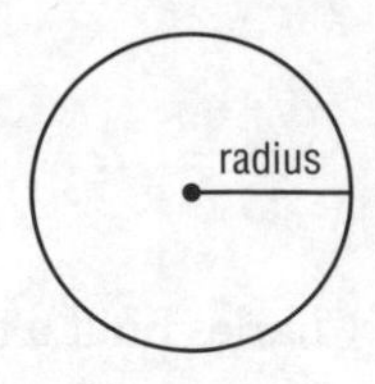


EXAMPLE Find the area of a circular hole in the floor with a radius of 2 inches.

$\frac{22}{7} \times \frac{2}{1} \times \frac{2}{1} = \frac{88}{7} = \mathbf{12\frac{4}{7}}$ **square inches**

Notice that the answer is in square units.

**Solve each problem. Be sure to label each answer in square units.**

1. Find the area of a circle with a radius of 7 feet. Use $\pi = \frac{22}{7}$.

2. If the radius of a circular pool is 10 feet, what is the area of the bottom of the pool? Use $\pi = 3.14$.

3. Mrs. Twyman wants to cover the top of a round table with square tiles that are each 1-inch square. How many tiles will she need if the radius of her table is 14 inches? Use $\pi = 3.14$.

4. If the tiles for the table in problem 3 cost $0.24 each, how much will Mrs. Twyman pay for the tiles to cover her table?

5. Find the area of a circular play yard with a radius of 20 feet that Norman built for his dog. Use $\pi = 3.14$.

**6.** To the nearest square foot, what is the area of a circular tablecloth with a radius of 4 feet? Use $\pi = 3.14$.

**7.** How many square yards of dancing space are there on a round dance floor with a radius of $3\frac{1}{2}$ yards? Use $\pi = \frac{22}{7}$.

**8.** If flooring costs $45 per square yard, how much will the flooring cost to cover the dance floor?

**9.** A circular vent in the wall of Mr. Henry's attic has a radius of one-half foot. How much material does he need to cover the vent? Use $\pi = \frac{22}{7}$.

**10.** The diagram shows the plan of a granite bathroom countertop with a hole for a sink. What is the area of the rectangular countertop?

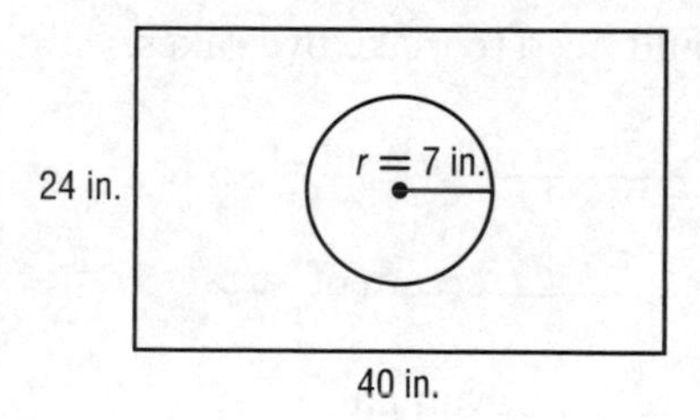


**11.** What is the area of the circular hole in the countertop? Use $\pi = \frac{22}{7}$.

**12.** After the hole is cut for the sink, what is the area of the remaining granite?

# Changing a Recipe

**Use the following list of ingredients needed for making a cake to answer the questions on this page.**

| | |
|---|---|
| 1 cup of flour | $1\frac{1}{3}$ cups of egg whites |
| $1\frac{1}{2}$ cups of sugar | $1\frac{2}{3}$ teaspoons of cream of tartar |
| $\frac{1}{4}$ teaspoon of salt | $1\frac{1}{4}$ teaspoons of vanilla |

**1.** Fill in the amount of each ingredient you would need to make two cakes.

| | |
|---|---|
| ________ flour | ________ egg whites |
| ________ sugar | ________ cream of tartar |
| ________ salt | ________ vanilla |

**2.** Fill in the amount of each ingredient you would need to make a smaller cake that is one-half the size of the cake in the recipe.

| | |
|---|---|
| ________ flour | ________ egg whites |
| ________ sugar | ________ cream of tartar |
| ________ salt | ________ vanilla |

**3.** Fill in the amount of each ingredient you would need to make five cakes.

| | |
|---|---|
| ________ flour | ________ egg whites |
| ________ sugar | ________ cream of tartar |
| ________ salt | ________ vanilla |

**4.** Joe decided to make a larger cake than the original recipe called for. He decided to use $1\frac{1}{2}$ times each of the original ingredients. Fill in the amounts he used.

| | |
|---|---|
| ________ flour | ________ egg whites |
| ________ sugar | ________ cream of tartar |
| ________ salt | ________ vanilla |

**5.** To solve problem 4, you can multiply the amount of each ingredient in problem 2 by what number?

# Converting Units of Measurement

In the United States we use a system of measurements called **standard** or **customary units.** The system includes familiar units such as feet, pounds, and gallons. Much of the rest of the world uses the **metric system** in which the basic unit length is the meter, the basic unit of weight is the kilogram, and the basic unit of liquid measure is the liter.

The two tables below include units of measurement of distance and area. Distance is a measurement along a straight line or a curve. Area is a measurement of the amount of surface on a flat figure. Notice that the list of equivalent distances includes two approximate equivalents. 1 mile is approximately equal to (≈) 1.61 kilometers, and 1 kilometer is approximately equal to 0.62 mile.

| Distance | Area |
|---|---|
| 1 foot = 12 inches | 1 square foot = 144 square inches |
| 1 yard = 3 feet | 1 square yard = 9 square feet |
| 1 mile = 5,280 feet | 1 acre = 43,560 square feet |
| 1 mile ≈ 1.61 kilometers | |
| 1 inch = 2.54 centimeters | |
| 1 foot = 0.3048 meter | |
| 1 meter = 1,000 millimeters | |
| 1 meter = 100 centimeters | |
| 1 kilometer = 1,000 meters | |
| 1 kilometer ≈ 0.62 mile | |

To change from a larger unit to a smaller unit, *multiply.*

EXAMPLE 1 **A table is 6 feet long. What is the length of the table in inches?**

According to the table, 1 foot = 12 inches. Multiply the length by 12.

$6 \times 12 =$ **72 inches**

To change from a smaller unit to a larger unit, *divide.*

EXAMPLE 2 **A rug has an area of 48 square feet. What is the area in square yards?**

According to the table, 1 square yard = 9 square feet. Divide 48 by 9.

$48 \div 9 = 5\frac{3}{9} = \mathbf{5\frac{1}{3}}$ **square yards**

EXAMPLE 3 **A board is 8 inches wide. Find the width in feet.**

1 foot = 12 inches. Divide 8 by 12.

$8 \div 12 = \frac{8}{12} = \mathbf{\frac{2}{3}}$ **foot**

**Use the tables on the previous page to solve each problem.**

1. The driveway to Mr. and Mrs. Walker's house is 126 feet long. What is the length of the driveway in yards?

2. The area of Mrs. Santiago's apartment is 750 square feet. What is the area in square yards?

3. The distance from Buffalo to Chicago is 540 miles. Find the distance in kilometers to the nearest *ten*.

4. A 15-inch pipe is how long in centimeters? Find the answer to the nearest *unit*.

5. Fred's desk is 1.5 meters long. What is the length in centimeters?

6. A real estate agent told Catherine that the property she wants to buy has an area of 10,890 square feet. What is the area in acres?

7. The distance from New York to London is 3,442 miles. To the nearest *unit*, what is the distance in kilometers?

8. A typical sheet of typing paper measures 11 inches by $8\frac{1}{2}$ inches. Find both of these measurements to the nearest *tenth* of a centimeter.

9. Stan runs a business burying industrial cable and charges by the mile. A client needs 2,640 feet of cable buried. How many miles will Stan charge for?

10. The ceilings in Mrs. Emerson's house are $9\frac{1}{2}$ feet high. What is the height of the ceiling in inches?

11. Jerome is 6 feet 4 inches tall. To the nearest *unit*, what is his height in centimeters?

The two tables below include units of volume and weight/mass. Volume is a measure of the amount of space occupied by a three-dimensional object. Weight and mass are measures of the heaviness of an object. Notice that the tables include several approximate equivalents. 1 liter is approximately 0.264 gallon, and 1 kilogram is approximately equal to 2.2 pounds.

| Volume | Weight/Mass |
| --- | --- |
| 1 cup = 8 fluid ounces | 1 ounce ≈ 28.350 grams |
| 1 quart = 4 cups | 1 pound = 16 ounces |
| 1 gallon = 4 quarts | 1 pound ≈ 453.592 grams |
| 1 gallon = 231 cubic inches | 1 milligram = 0.001 grams |
| 1 liter ≈ 0.264 gallon | 1 kilogram = 1,000 grams |
| 1 cubic foot = 1,728 cubic inches | 1 kilogram ≈ 2.2 pounds |
| 1 cubic yard = 27 cubic feet | 1 ton = 2,000 pounds |
| 1 board foot = 1 inch by 23 inches by 23 inches | |

**Use the tables above to solve each problem.**

**12.** As an administrative assistant, Marcy has to ship a package weighing 12 ounces. The mailroom requires a weight measured in pounds. Find the weight of the envelope in pounds.

**13.** Steve needs five cubic yards of topsoil for a landscaping job. His supplier sells topsoil by the cubic foot. How many cubic feet should Steve order?

**14.** The sign in an elevator states that maximum allowable load is 3,500 pounds. Find the weight in tons that the elevator can safely carry.

**15.** Silvia made $6\frac{1}{2}$ gallons of lemonade for her company's annual picnic. How many quarts of lemonade did she make?

**16.** At birth Max weighed 3.5 kilograms. To the nearest tenth, what was his birth weight in pounds?

**17.** To make soup, Luca used 6 cups of vegetable stock. How many quarts of stock did he use?

**18.** What is the weight in grams of a box of books that weighs 4.2 kilograms?

# Finding Interest for One Year

Interest is money that money earns. On a loan, **interest** is the payment you must make for using the lender's money. In a savings account, interest is the money the bank pays you for using your money.

The formula for finding interest is $i = prt$.

$i$ is the *interest* in dollars.

$p$ is the *principal*, the money borrowed or saved.

$r$ is the percent *rate*, which can be written as either a fraction or a decimal.

$t$ is the *time* in years.

The formula is read as *interest is equal to the principal times the rate times the time.*

EXAMPLE 1 **Find the interest on $500 at 8% annual interest for one year.**

$$\frac{\overset{5}{\cancel{500}}}{1} \times \frac{8}{\underset{1}{\cancel{100}}} \times 1 = \mathbf{\$40}$$

Notice that the percent was changed to a fraction with a denominator of 100.

EXAMPLE 2 **Find the interest on $600 at $4\frac{1}{2}$% annual interest for one year.**

**STEP 1** Change the complex percent to a fraction with a denominator of 100. $\frac{4\frac{1}{2}}{100}$

**STEP 2** Write the complex fraction as a division problem and simplify. $4\frac{1}{2} \div 100 = \frac{9}{2} \times \frac{1}{100}$

**STEP 3** Multiply 600 by the simplified percent. $\frac{9}{2} \times \frac{1}{\underset{1}{\cancel{100}}} \times \frac{\overset{6}{\cancel{600}}}{1} = \frac{54}{2} = \mathbf{\$27}$

EXAMPLE 3 **Find the interest on $400 at 8.5% for one year.**

$$\begin{array}{r} \$400 \\ \times\ 0.085 \\ \hline 2\ 000 \\ 32\ 00 \\ \hline 34\ 00\cancel{0} \\ \times\ \ 1 \\ \hline \mathbf{\$34.00} \end{array}$$

Write 8.5% as a decimal (0.085).

**Solve each problem.**

1. Manny borrowed $800 at 10% annual interest. How much interest did he owe in one year?

2. Deborah borrowed $450 at 15% annual interest. How much interest did she owe in one year?

3. Frank put $200 in his savings account for one year. If his money earned $2\frac{1}{2}$% annual interest, how much interest did he get in a year?

4. Diane borrowed $5,000 in order to finish a year in school. If she had to pay 6.7% interest on the loan, how much interest did she owe after one year?

5. Ellen saved $3,600 for a year. How much interest did she earn if her bank pays an annual interest rate of $2\frac{1}{4}$%?

6. How much interest did Daniel owe on a $1,500 loan in one year if his bank charged him 11.5% interest?

7. José had to pay 8.4% interest on a $950 loan. How much interest did he owe in one year?

8. Madge kept $1,200 in a certificate of deposit (CD) for a year. How much interest did she earn if her CD paid $4\frac{3}{4}$% annual interest?

# Finding Interest for Less than One Year

Interest rates are usually given for one year. If you want to find interest for less than one year, write the number of months over 12 (the number of months in one year). Use this fraction for $t$ in the formula $i = prt$.

EXAMPLE **Find the interest on $500 at 8% annual interest for nine months.**

STEP 1 $9 \text{ months} = \frac{9 \text{ months}}{12 \text{ months per year}} = \frac{3}{4} \text{ year}$

STEP 2 $\frac{\overset{5}{\cancel{500}}}{1} \times \frac{\overset{2}{\cancel{8}}}{\underset{1}{\cancel{100}}} \times \frac{3}{\underset{1}{\cancel{4}}} = \mathbf{\$30}$

**Solve each problem.**

1. Find the interest on $900 at 6% annual interest for four months.

2. Mary's bank pays 2.5% annual interest on savings accounts. How much interest will Mary make on $1,000 in six months?

3. David's charge account costs him 18% in annual interest. How much interest would he owe on $700 in 10 months?

4. Jane kept $270 in her savings account for 8 months. How much interest will she get if her bank pays 2% annual interest?

5. Fred was charged 9.2% annual interest on a $1,200 loan. How much interest did he owe if he paid the loan back in two months?

6. For Fred, from the last problem, what total amount did he have to pay at the end of two months?

# Finding Interest for More than One Year

If you want to find interest for more than one year, write the total number of months over 12 (the number of months in one year). Use this improper fraction for $t$ in the formula $i = prt$.

EXAMPLE **Find the interest on $600 at 5% annual interest for one year and 6 months.**

STEP 1 $1 \text{ year and } 6 \text{ months} = \frac{18 \text{ months}}{12 \text{ months per year}} = \frac{3}{2} \text{ years}$

STEP 2 $\frac{\overset{3}{\cancel{6}}00}{1} \times \frac{5}{\underset{1}{\cancel{100}}} \times \frac{3}{\underset{1}{\cancel{2}}} = \mathbf{\$45}$

**Solve each problem.**

1. Find the interest on $800 at 9% annual interest for two years and three months.

2. Carlos borrowed $1,500 at 11% annual interest. He paid the loan back in one year and eight months. How much interest did he pay?

3. Sally kept $400 in her bank account for one year and nine months. How much interest did she make if her bank pays $2\frac{1}{2}$% annual interest?

4. Mr. Clay paid back his $2,000 car loan two years and six months after he borrowed the money. If he was charged 10.5% annual interest, how much interest did he pay?

5. What total amount including interest did Mr. Clay, in problem 4, have to pay for his car loan?

6. Jack's bank pays $2\frac{1}{4}$% annual interest on Jack's account. How much interest will Jack make on a $3,000 deposit that he keeps in the bank for one year and four months?

# Finding Compound Interest

In most savings accounts, money earns **compound interest.** This means that your balance (the principal) changes by the addition of interest on a regular time basis such as quarterly (every three months), monthly, or even daily.

EXAMPLE **Joe deposited $500 in his savings account. How much money will he have in the account at the end of a year if he gets 4% annual interest and the interest is compounded (added to his balance) quarterly?**

**STEP 1** First quarter: $\frac{\overset{5}{\cancel{500}}}{1} \times \frac{\overset{1}{\cancel{4}}}{\underset{1}{\cancel{100}}} \times \frac{1}{\underset{1}{\cancel{4}}} = \$5.00$

**Amount in account:** $500 + $5 = **$505.00**

**STEP 2** Second quarter: Joe gets interest on the new balance, $505.00.

$$\frac{\overset{101}{\cancel{505}}}{1} \times \frac{\overset{1}{\cancel{4}}}{\underset{20}{\cancel{100}}} \times \frac{1}{\underset{1}{\cancel{4}}} = \frac{101}{20} = \$5.05$$

**Amount in account:** $505 + $5.05 = **$510.05**

**STEP 3** Third quarter: The interest is calculated on $510.05.

$$\frac{\overset{102.01}{\cancel{510.05}}}{1} \times \frac{\overset{1}{\cancel{4}}}{\underset{20}{\cancel{100}}} \times \frac{1}{\underset{1}{\cancel{4}}} = \frac{102.01}{20} = \$5.10 \text{ (rounded to the nearest cent)}$$

**Amount in account:** $510.05 + $5.10 = **$515.15**

**STEP 4** Fourth quarter: The interest is calculated on $515.15.

$$\frac{\overset{103.03}{\cancel{515.15}}}{1} \times \frac{\overset{1}{\cancel{4}}}{\underset{20}{\cancel{100}}} \times \frac{1}{\underset{1}{\cancel{4}}} = \frac{103.03}{20} = \$5.15 \text{ (rounded to the nearest cent)}$$

**Amount in account:** $515.15 + $5.15 = **$520.30**

Notice that every quarter the new principal was used to calculate the amount of interest.

**Solve each problem. Round each answer to the nearest cent.**

Luis put $800 in his savings account which pays 6% annual interest. How much will he have in his account in a year if the interest is compounded quarterly?

**1. a.** first quarter interest **b.** first quarter principal

**2. a.** second quarter interest **b.** second quarter principal

**3. a.** third quarter interest **b.** third quarter principal

**4. a.** fourth quarter interest **b.** fourth quarter principal

To pay off her high-interest credit card debt, Denise borrowed $1,200 at 8% annual interest. How much will she owe in a year if the interest is compounded quarterly?

**5. a.** first quarter interest **b.** first quarter principal

**6. a.** second quarter interest **b.** second quarter principal

**7. a.** third quarter interest **b.** third quarter principal

**8. a.** fourth quarter interest **b.** fourth quarter principal

To cover college expenses Jake borrowed $20,000 from his older brother Jed. They agreed that Jake would pay a low 5% interest, compounded annually for four years. How much will Jake owe his brother at the end of four years?

**9. a.** first year interest **b.** first year principal

**10. a.** second year interest **b.** second year principal

**11. a.** third year interest **b.** third year principal

**12. a.** fourth year interest **b.** fourth year principal

# Finding the Percent Saved at a Sale

To find what percent you save by buying an item on sale, put the amount saved over the original price, reduce, and change the fraction into a percent.

EXAMPLE **All the prints that fit . . . on sale!** Long-sleeved boys' shirts in assorted prints. 100% cotton, sizes S–M–L.
reg. $15 ............................ **sale $10**

What percent of the original price can you save by buying the shirt on sale?

STEP 1 The regular price is $15. The sale price is $10. The amount saved is $15 – $10 = $5.

STEP 2 $\frac{\text{savings}}{\text{original}} = \frac{\$5}{\$15} = \frac{1}{3} = \mathbf{33\frac{1}{3}\%}$

**Find the percent of savings for each item shown in this advertisement.**

**STARR'S DISCOUNT STORE**

**Let Us Make You a Starr!**

**Gigantic Savings—Storewide Clearance Sale**

**Wizard® canister vacuum.** 2.3 hp (peak-rated by manufacturer); automatic cord rewind; tools, more! Add $5 delivery.
orig. $120 ...................... **sale $90**

**Fashion's most popular pre-washed denim fabric.** 45" wide.
reg. $12 ...................... **sale $6 yd**

**Men's hooded fleece jackets.** Zip or button-front. S-M-L-XL.
reg. $45 ......................... **sale $30**

**Misses' cotton crewneck T-shirt for skirts or jeans!** Stripe combinations of blue, green, brown. S-M-L.
reg. $15 .......................... **sale $9**

**Men's nylon warm-up suits.** Buy separates or set. Blue or bone. Team name on jacket. Sizes S-M-L-XL.
reg. $36 each ........ **sale $27 each**

**Men's sleeveless print sport shirts.** Splashy patterns, geometrics; assorted colors. Cotton, polyester/cotton or nylon. S-M-L-XL.
orig. $24 ........................ **sale $16**

**Girls' pants.** Daisy print. Sizes 8, 10, 12, 14, 16.
orig. $20 ........................ **sale $13**

**Men's cotton shorts.** Sizes 32–40.
reg. $18 ......................... **sale $15**

**Boys' canvas slip-on shoes.** Assorted colors; rubber sole. Sizes 1–11.
reg. $12.50 ..................... **sale $9**

**Geometric-design blankets.** 100% acrylic; machine wash, tumble dry. Brown/blue reverses to solid blue.
Twin, orig. $35 .............. **sale $28**

Full, orig. $45 ............... **sale $36**

1. Wizard canister vacuum

2. Pre-washed denim fabric

3. Men's hooded fleece jacket

4. Misses' cotton T-shirt

5. Men's nylon warm-up suit

6. Men's sleeveless sport shirt

7. Girls' pants

8. Men's cotton shorts

9. Boys' canvas shoes

10. Twin-size blanket

11. Serena shopped at Starr's Discount to buy clothes for her family. She purchased a hooded fleece jacket for her husband, three misses' T-shirts, two pairs of girls' pants, and one pair of boys' canvas slip-on shoes. Including a 6.5% sales tax, what was the total cost of her purchases?

12. You are an employee at Starr's Discount, and you decided to take advantage of the sale. You purchased a Wizard canister vacuum, two pairs of men's nylon warm-up pants, and two men's sleeveless sport shirts. As an employee you get an extra 10% discount on every item you buy. Including the 6.5% sales tax, what the total price of your purchases?

# Buying Furniture on Sale

When Deborah Robinson went into the Friendly Furniture Store, she saw these signs.

| EVERY LAMP<br>25% OFF | ALL BABY FURNITURE<br>$33\frac{1}{3}$% OFF |
|---|---|

Deborah decided to buy a pair of lamps that had originally sold for $78.00 each and a high chair that originally cost $49.95.

**Use the information above to answer the following questions.**

1. How much money could she save on each lamp?

2. What was the sale price of each lamp?

3. How much money could she save on the high chair?

4. What was the sale price of the high chair?

5. What was the total price for the two lamps and the high chair?

6. In Deborah's state there is an 8% sales tax. How much did she owe for her furniture including tax?

7. Since Deborah doesn't drive, she had to pay $25 to have her purchases delivered. Find the total price she paid including the delivery charge.

# Calculating Rates

Think about the phrase "miles per gallon." The phrase describes how fuel efficient a car is. A car that gets only 13 miles per gallon is far less efficient (and more expensive to use) than a car that gets 32 miles per gallon.

A *rate unit* describes how a rate is measured. Miles per gallon means miles for every one gallon. Rates are often written like fractions with the slash (/). For example, a worker's wage in dollars per hour can be written $/hour.

To calculate a rate, divide the first number by the amount that follows the word *per* or *for each*.

**Calculate each of the following rates and give each answer the correct label, such as miles per gallon.**

1. On the street where Melanie lives, there are 19 families and a total of 46 children. To the nearest tenth, what is the number of children per family on Melanie's street?

2. Alex works as a roofer. He charged a customer $1,000 plus the cost of materials to put a new roof on their summer cottage. If he worked 46 hours on the project, how much did he earn in dollars per hour?

3. Silvia has to assign classrooms at the Oakdale School. 928 students are registered, and there are 35 classrooms at the school. To the nearest tenth, what will be the number of students per classroom?

4. Phil owns a commercial painting business. He has a contract to repaint the rooms in two motels. One motel has 38 rooms, and the other has 65 rooms. Phil has seven full-time painters. Find the number of rooms per person that each painter will have to paint to complete the job at the motels.

5. The 42 members of the Friends of the Library association want to sell 1,500 tickets for their annual raffle. How many tickets per member will the organization have to sell?

6. A plane flew 2,419 miles in 6.5 hours. What was the speed of the flight in miles per hour?

7. Sandra drove 475 miles on 22 gallons of gasoline. Calculate the number of miles per gallon that her car gets.

8. The employees in a post office processed 1,500 pieces of out-going mail in 9 hours. To the nearest whole number, how many pieces of mail were processed per hour?

# Comparing Food Prices: Unit Pricing

Similar food items come in different package sizes. It is not always easy to tell which package is the best buy. Suppose you find two different brands of peas, one in a 16-ounce can priced at 72¢ and another in a 10-ounce can priced at 55¢. To find the best buy, calculate the price per ounce of peas in each can. The word *per* suggests division. Finding this price is called **unit pricing.**

EXAMPLE Which of these two brands of peas is cheaper?

16-ounce can of Brand X peas for 72¢

10-ounce can of Brand Y peas for 55¢

The cost per ounce for Brand X is

$$\begin{array}{r} 4\frac{8}{16} \\ 16\overline{)72} \\ \underline{64} \\ 8 \end{array} = \mathbf{4\frac{1}{2}\text{¢ per ounce}}$$

The cost per ounce for Brand Y is

$$\begin{array}{r} 5\frac{5}{10} \\ 10\overline{)55} \\ \underline{50} \\ 5 \end{array} = \mathbf{5\frac{1}{2}\text{¢ per ounce}}$$

ANSWER: **Brand X is cheaper.**

**Use the advertising posters below to answer the questions on the next page.**

**FRED'S FOODS**

Margarine 2 8-oz cups $2.29

48 size, Indian River seedless grapefruit 7 for $2.49

orange juice 1/2-gallon cont. $2.59

gourmet pasta sauce 16-oz jar $2.99

**GERT'S GROCERIES**

Margarine 3 8-oz cups $3.09

INDIAN RIVER SEEDLESS GRAPEFRUIT (48 SIZE) 6 for $2.29

Orange Juice 64-OZ BTL. $2.49

gourmet pasta sauce 2 24-oz jars $9.00

**SAM'S STORE**

Margarine 1-lb pkg. $1.99

FLORIDA SEEDLESS Grapefruit 4 for $1.29

Orange Juice 3 1-qt conts. $3.95

gourmet pasta sauce 24-oz. jar $5.85

**1.** What was the price of 8 ounces of margarine

**a.** at Fred's Foods?

**b.** at Gert's Groceries?

**c.** at Sam's Store? (1 pound = 16 ounces)

**d.** Which store had the least expensive margarine?

**2.** What was the price of one seedless grapefruit

**a.** at Fred's Foods?

**b.** at Gert's Groceries?

**c.** at Sam's Store?

**d.** Which store had the least expensive grapefruit?

**3.** What was the price of one quart (32 ounces) of orange juice

**a.** at Fred's Foods?

**b.** at Gert's Groceries?

**c.** at Sam's Store?

**d.** Which store had the least expensive orange juice?

**4.** What was the price of 8 ounces of gourmet pasta sauce

**a.** at Fred's Foods?

**b.** at Gert's Groceries?

**c** at Sam's Store?

**d.** Which store had the least expensive gourmet pasta sauce?

# Using a Tax Rate Schedule

An important part of filling out income tax forms is using a tax rate schedule. Tax rate schedules let you know how much money you have to pay in taxes based on how much money you made during the year. The tax rate schedule on this page is for single tax payers. This schedule is used to figure the amount owed to the federal government. The amount is based on the taxable income entered on line 5 of tax form 1040. Both tax rate schedules and tax forms can be obtained from the Internal Revenue Service.

**Schedule X—Single**

| **If line 5 is:** Over– | *But not over–* | **The tax is:** | *of the amount over–* |
|---|---|---|---|
| $0 | $8,375 | $0 + 10% | $0 |
| 8,375 | 34,000 | $837.50 + 15% | 8,375 |
| 34,000 | 82,400 | 4,681.25 + 25% | 34,000 |
| 82,400 | 171,850 | 16,781.25 + 28% | 82,400 |
| 171,850 | 373,650 | 41,827.25 + 33% | 171,850 |
| 373,650 | and greater | 108,421.25 + 35% | 373,650 |

EXAMPLE **How much tax does a single tax payer owe if the amount on line 5 is $31,000?**

STEP 1 Read the tax rate schedule from left to right. Since $31,000 is between $8,375 and $34,000, the tax is $837.50 plus 15% of the amount over $8,375.

STEP 2 Find the amount over $8,375.

$$\begin{array}{r} \$31{,}000 \\ -\ \ 8{,}375 \\ \hline \$22{,}625 \end{array}$$

STEP 3 Find 15% of 22,625.

$$\begin{array}{r} \$22{,}625 \\ \times\ \ \ 0.15 \\ \hline 1131\ 25 \\ 2262\ 5\ \ \\ \hline \$3{,}393.75 \end{array}$$

STEP 4 Add.

$$\begin{array}{r} \$837.50 \\ +\ 3{,}393.75 \\ \hline \mathbf{\$4{,}231.25} \end{array}$$ **tax owed**

**Use the schedule on page 162 to solve problems 1 to 5.**

1. If line 5 is $16,240, how much tax is due?

2. Linda is a plant supervisor. If line 5 on her 1040 form is $58,910, how much tax does she owe?

3. Jed is a student with a part-time job. If the amount on line 5 of his 1040 form is $5,910, how much tax does he owe?

4. Cheryl is a real estate broker. The amount on line 5 of her 1040 is $85,600. How much tax does she owe?

5. Kate is an executive in a high-tech company. If line 5 on her 1040 form is $176,000, how much tax does she owe?

Schedule Z is the tax rate schedule for heads of household. The **head of household** is the person who brings in the money or most of the money on which a family lives.

**Schedule Z—Head of household**

| **If line 5 is:** Over– | But not over– | **The tax is:** | of the amount over– |
|---|---|---|---|
| $0 | $11,950 | $0 + 10% | $0 |
| 11,950 | 45,550 | $1,195.00 + 15% | 11,950 |
| 45,550 | 117,650 | 6,235.00 + 25% | 45,550 |
| 117,650 | 190,550 | 24,260.00 + 28% | 117,650 |
| 190,550 | 373,650 | 44,672.00 + 33% | 190,550 |
| 373,650 | and greater | 105,095.00 + 35% | 373,650 |

**Use Schedule Z to answer questions 6 to 10.**

6. If line 5 is $31,000, how much tax is due?

7. If line 5 on Marco's 1040 is $47,330, how much tax does he owe?

8. Jill sells medical equipment. If line 5 on her 1040 is $121,900, how much federal tax does she have to pay?

9. If line 5 on Mr. Quinter's 1040 form is $68,300, how much tax does he owe?

10. Charlaine worked only half the year. The amount on line 5 of her 1040 form is $24,290. How much tax does she owe?

# Filling Out a Wage and Tax Statement

Tax payers in the United States have to pay FICA (payroll) tax. FICA stands for the Federal Insurance Contributions Act. There are two parts to FICA. One payment is for social security, and the other is for Medicare.

Social security tax is 6.2% of a worker's adjusted gross income.

Medicare tax is 1.45% of a worker's adjusted gross income.

**Jane's adjusted gross salary one year was $40,000. Calculate the amount of social security tax and the amount of Medicare tax that was withheld from her wages.**

Social security tax = 6.2% of $40,000 = 0.062 × $40,000 = $2,480.

Medicare tax = 1.45% of $40,000 = 0.0145 × $40,000 = $580.

---

**For each problem fill in amounts for wages, tips, other compensation; for social security tax withheld; and for Medicare tax withheld. Round all amounts to the nearest dollar.**

**1.** Peter Cheung's adjusted gross annual salary was $35,000.

Wages, tips, other compensation ______________

Social security tax withheld ______________

Medicare tax withheld ______________

**2.** Jeanne made $16,500 in adjusted gross wages as a teacher's aid and an additional $12,800 working in the administration office of her school.

Wages, tips, other compensation ______________

Social security tax withheld ______________

Medicare tax withheld ______________

3. Blanca earned $34,300 as a restaurant manager and an additional $4,160 in tips by filling in for an absent waiter.

Wages, tips, other compensation ________________

Social security tax withheld ________________

Medicare tax withheld ________________

4. Jim works for his family's landscaping business. One year he made $36,230 in wages and an additional $9,850 filling in for his mother in the office.

Wages, tips, other compensation ________________

Social security tax withheld ________________

Medicare tax withheld ________________

5. Ed made $42,000 in salary and $17,900 in commissions at his job as a car salesman.

Wages, tips, other compensation ________________

Social security tax withheld ________________

Medicare tax withheld ________________

6. Janet works for a trucking company. One year she made $20,400 as a radio dispatcher. She also made $6,750 filling in for a driver, and she made an additional $5,700 as a bookkeeper in the payroll department.

Wages, tips, other compensation ________________

Social security tax withheld ________________

Medicare tax withheld ________________

# Working with a Budget

The budget below shows the way that the four members of the Johnson family spend their money.

1. The income for the Johnson family comes from Mr. Johnson's yearly salary of \$50,000. 28% of his salary is withheld for taxes and social security. How much money is withheld for taxes and social security in a year?

2. What is Mr. Johnson's yearly take-home pay?

**Use the budget below and Mr. Johnson's take-home pay to fill in the amount budgeted for each expense.**

3.

| **Johnson Family Yearly Budget** | | |
|---|---|---|
| **Expenses** | **% Budgeted** | **Amount Budgeted** |
| Rent | 20% | |
| Food | 31% | |
| Utilities | 5% | |
| Entertainment | 6% | |
| Clothes | 10% | |
| Medical | 12% | |
| Savings | 10% | |
| Odds and Ends | 6% | |

4. How much do the Johnsons spend each month to rent their one-bedroom apartment?

5. Next year Mrs. Johnson plans to get a part-time job as a teacher's assistant. The Johnsons expect their income to increase by 15%. If they continue to spend the same percent on rent, how much monthly rent can they afford next year?

The budget below shows the yearly expenses of Gus and Elena Kaligas.

**Use the budget below to answer the next questions.**

**Kaligas Family Yearly Budget**

| **Expenses** | **Amount Budgeted** | **% Budgeted** |
| --- | --- | --- |
| Housing | $20,160 | |
| Transportation | $10,080 | |
| Food | $ 8,820 | |
| Clothing | $ 3,150 | |
| Health Care | $ 3,780 | |
| Insurance and Pensions | $ 5,670 | |
| Entertainment | $ 3,780 | |
| Other | $ 7,560 | |

**6.** Find the Kaligas family's total expenses for the year.

**7.** What percent of their total yearly budget did they spend for each category?

**8.** Together housing, transportation, and food represent what percent of the Kaligas family's budget?

**9.** Gus and Elena decided to save one-third of the amount budgeted for "other." How much will they save in one year?

**10.** What are the Kaligas family's average monthly expenses for housing?

# Buying on an Installment Plan

Buying on an installment plan is a way to make partial payments for an item. However, by paying in regular installments, a customer can end up paying much more than the list price of an item. The problems that follow show how different cash prices and installment prices can be.

**For each problem, use the given information to answer the questions.**

1. A lawn mower marked $159 can be bought for 10% down and $15 a week for 12 weeks.

   **a.** What is the cash price of the lawn mower including 8% sales tax?

   **b.** How much does the lawn mower cost on the installment plans?

   **c.** How much more does it cost to buy the lawn mower on the installment plan than it does to pay cash?

2. An upright vacuum cleaner is priced at $89. It can be bought for no payment down and $11 a month for 1 year.

   **a.** What is the cash price of the vacuum cleaner including 7% sales tax?

   **b.** How much does the vacuum cleaner cost on the installment plan?

   **c.** How much more does it cost to buy the vacuum cleaner on the installment plan than it does to pay cash?

3. A kitchen table and chair set marked $395 can be bought for 15% down and $35 a month for 1 year.

   **a.** What is the cash price of the furniture including 6% sales tax?

   **b.** How much does the furniture cost on the installment plan?

   **c.** How much more does it cost to buy the furniture on the installment plan than it does to pay cash?

4. A color television marked $459.00 can be bought for 8% down and $10.50 a week for 1 year.

   a. What is the cash price of the television including 5% sales tax?

   b. How much does the television cost on the installment plan?

   c. How much more does it cost to buy the television on the installment plan than it does to pay cash?

5. A mattress and box-spring set is priced at $289.95. They can be bought for 12% down and $14.80 a week for 22 weeks.

   a. What is the cash price of the mattress and box-spring set including 6% sales tax?

   b. How much does the mattress and box-spring set cost on the installment plan?

   c. How much more does it cost to buy the mattress and box-spring set on the installment plan than it does to pay cash?

6. An air conditioner marked $229 can be bought for 16% down and $25 per month for 1 year.

   a. What is the cash price of the air conditioner including 5% sales tax?

   b. How much does the air conditioner cost on the installment plan?

   c. How much more does it cost to buy the air conditioner on the installment plan than it does to pay cash?

# Posttest B

This review test has a multiple-choice format much like the GED and other standardized tests. Take your time and work each problem carefully. Circle the correct answer to each problem. When you finish, check your answers at the back of the book.

**1.** Reduce $\frac{36}{54}$ to lowest terms.

**a.** $\frac{3}{7}$ **b.** $\frac{3}{5}$ **c.** $\frac{2}{3}$ **d.** $\frac{2}{5}$

**2.** $5\frac{3}{16} + 4\frac{2}{3} + 6\frac{1}{2} =$

**a.** $16\frac{17}{48}$ **b.** $16\frac{9}{16}$ **c.** $15\frac{7}{48}$ **d.** $15\frac{3}{16}$

**3.** Which of the following best describes the sum of $\frac{5}{9} + \frac{8}{15}$?

**a.** greater than 1 **b.** less than 1 **c.** equal to 1 **d.** less than 0

**4.** Find the combined weight of a motor that weighs $48\frac{5}{16}$ pounds and a wooden packing crate that weighs $3\frac{3}{4}$ pounds.

**a.** $49\frac{7}{16}$ pounds **b.** $60\frac{1}{16}$ pounds **c.** $51\frac{7}{16}$ pounds **d.** $52\frac{1}{16}$ pounds

**5.** $8\frac{1}{3} - 3\frac{4}{5} =$

**a.** $5\frac{7}{15}$ **b.** $5\frac{2}{15}$ **c.** $4\frac{8}{15}$ **d.** $4\frac{4}{15}$

**6.** A door opening is 68 inches high, and Jason is $62\frac{1}{2}$ inches tall. How much taller is the door frame than Jason?

**a.** $4\frac{1}{2}$ inches **b.** $5\frac{1}{2}$ inches **c.** $6\frac{1}{2}$ inches **d.** $7\frac{1}{2}$ inches

**7.** $\frac{9}{16} \times \frac{2}{3} =$

**a.** $\frac{7}{16}$ **b.** $\frac{3}{16}$ **c.** $\frac{7}{8}$ **d.** $\frac{3}{8}$

**8.** $2\frac{1}{4} \times \frac{2}{3} \times 1\frac{5}{6} =$

**a.** $3\frac{1}{4}$  **b.** $2\frac{3}{4}$  **c.** $2\frac{5}{6}$  **d.** $1\frac{5}{6}$

**9.** In the problem $1\frac{7}{8} \times 6\frac{5}{6}$, first round each number to the nearest whole number. Then multiply.

**a.** 6  **b.** 7  **c.** 12  **d.** 14

**10.** What is the total weight of three packages of grass seed, if each weighs $1\frac{3}{4}$ pounds?

**a.** $5\frac{1}{4}$ pounds  **b.** $4\frac{3}{4}$ pounds  **c.** $4\frac{1}{4}$ pounds  **d.** $3\frac{3}{4}$ pounds

**11.** $3\frac{1}{5} \div 4 =$

**a.** $\frac{3}{4}$  **b.** $\frac{3}{5}$  **c.** $\frac{4}{5}$  **d.** $1\frac{1}{5}$

**12.** $2\frac{2}{3} \div 1\frac{1}{9} =$

**a.** $1\frac{3}{5}$  **b.** $1\frac{4}{5}$  **c.** $2\frac{1}{5}$  **d.** $2\frac{2}{5}$

**13.** How many bags, each containing $1\frac{3}{4}$ pounds of topsoil, can be filled from a pile of soil that weighs 350 pounds?

**a.** 130 bags  **b.** 150 bags  **c.** 175 bags  **d.** 200 bags

**14.** Laurie is getting signatures on a petition. So far she has 240 signatures. This is $\frac{3}{4}$ of her goal. How many signatures is she trying to get?

**a.** 320  **b.** 300  **c.** 280  **d.** 250

**15.** What is seven and fifty-one thousandths written as a mixed decimal?

**a.** 75.1  **b.** 7.51  **c.** 7.051  **d.** 7.0051

16. Write the following decimals in order from smallest to largest:
0.06, 0.064, 0.4, 0.004

a. 0.004, 0.4, 0.064, 0.06
b. 0.004, 0.06, 0.064, 0.4
c. 0.06, 0.4, 0.004, 0.064
d. 0.06, 0.064, 0.004, 0.4

17. $3.49 + 8 + 14.902 =$

a. 24.61
b. 24.402
c. 26.392
d. 26.41

18. In the year 2010, the population of the United States was 310.2 million. Experts think the population in 2050 will be 128.8 million more than in 2010. What is the population of the United States expected to be in 2050?

a. 442 million
b. 439 million
c. 428 million
d. 319 million

19. $32 - 19.786$

a. 19.754
b. 19.466
c. 12.786
d. 12.214

20. In the problem $2.749 - 0.1847$, first round each number to the nearest *hundredth.* Then subtract.

a. 2.57
b. 2.92
c. 4.4
d. 9.02

21. One yard is equal to 0.914 meter. A meter is how much longer than one yard?

a. 0.014 meter
b. 0.086 meter
c. 0.114 meter
d. 0.186 meter

22. $17.6 \times 3.09 =$

a. 54.384
b. 5.4384
c. 0.54384
d. 543.84

23. In the problem $60.8 \times 0.4$, first round 60.8 to the nearest *whole number.* Then multiply.

a. 240
b. 24
c. 24.4
d. 24.48

24. Find the cost of 1.7 pounds of cheese at a cost of $4.98 per pound.

a. $6.68
b. $7.47
c. $7.68
d. $8.47

**25.** $39.56 \div 9.2 =$

**a.** 0.23 **b.** 2.3 **c.** 4.3 **d.** 43

**26.** $6 \div 1.5 =$

**a.** 40 **b.** 4 **c.** 2.5 **d.** 0.25

**27.** Find the answer to $7.85 \div 1.2$ to the nearest *hundredth.*

**a.** 65.41 **b.** 15.29 **c.** 6.54 **d.** 0.65

**28.** Carla drove 144 miles in 2.5 hours. To the nearest unit, find her average speed in miles per hour.

**a.** 50 mph **b.** 52 mph **c.** 55 mph **d.** 58 mph

**29.** Write 2.6% as a decimal.

**a.** 2.6 **b.** 0.26 **c.** 0.026 **d.** 0.0026

**30.** Write $\frac{3}{7}$ as a percent.

**a.** $42\frac{6}{7}\%$ **b.** $39\frac{5}{7}\%$ **c.** $23\frac{1}{3}\%$ **d.** $4\frac{2}{7}\%$

**31.** Write $31\frac{1}{4}\%$ as a common fraction.

**a.** $\frac{5}{8}$ **b.** $\frac{5}{9}$ **c.** $\frac{5}{12}$ **d.** $\frac{5}{16}$

**32.** Find 42% of 700.

**a.** 2.94 **b.** 29.4 **c.** 294 **d.** 2,940

**33.** Find $87\frac{1}{2}\%$ of 240.

**a.** 21 **b.** 30 **c.** 180 **d.** 210

**34.** Which of the following is the same as $16\frac{2}{3}\%$ of 180?

**a.** $180 \div 2$ **b.** $180 \div 4$ **c.** $180 \div 6$ **d.** $180 \div 8$

**35.** 84 is what percent of 96?

**a.** $87\frac{1}{2}\%$ **b.** 84% **c.** 45% **d.** $62\frac{1}{2}\%$

**36.** One year Miguel taught a course in furnace maintenance at a community college. That year 24 students registered for the class. When the college offered the class the next year, 42 students registered. By what percent did the number of registrations increase?

**a.** 25% **b.** 51% **c.** 60% **d.** 75%

**37.** 65% of what number is 78?

**a.** 51 **b.** 90 **c.** 120 **d.** 186

**38.** At the Central School night program, 196 students said that they planned to go on to further study. They represent 70% of the total number of students. How many students are in the night program?

**a.** 296 **b.** 280 **c.** 226 **d.** 137

# POSTTEST B CHART

If you missed more than one problem on any group below, review the practice pages for those problems. If you had a passing score, redo any problem you missed.

| Problem Numbers | Skill Area | Practice Pages |
|---|---|---|
| 1, 2, 3, 4, 5, 6, 7, 8, 9, 10, 11, 12, 13, 14 | using fractions | 11–59 |
| 15, 16, 17, 18, 19, 20, 21, 22, 23, 24, 25, 26, 27, 28 | using decimals | 64–96 |
| 29, 30, 31, 32, 33, 34, 35, 36, 37, 38 | using percents | 101–124 |

# ANSWER KEY

**Pages 1–5, Pretest**

1. $\frac{3}{4}$
2. $8\frac{23}{24}$
3. greater than 1
4. $36\frac{1}{8}$ inches
5. $4\frac{19}{30}$
6. **b.** $9 - 7 =$
7. $2\frac{13}{16}$ pounds
8. $\frac{5}{18}$
9. 35
10. 15
11. 16 students
12. $\frac{1}{2}$
13. $2\frac{8}{9}$
14. $3\frac{1}{2}$ inches
15. 48 words
16. 12.016
17. 0.057, 0.075, 0.57, 0.7
18. 18.708
19. 1.625 inches
20. 1.703
21. 6.8
22. 4.8 square miles
23. 10.582
24. 72
25. $12.77
26. 3.8
27. 0.193
28. 200
29. 12.6
30. 42.9 pounds
31. 0.048
32. $41\frac{2}{3}\%$
33. $\frac{1}{16}$
34. 81
35. 6
36. **d.** $75 \div 5$
37. 14 items
38. 2,760 people
39. 75%
40. $62\frac{1}{2}\%$
41. 80%
42. 60%
43. 120
44. 130 seats
45. 48 streets

**Pages 8–10, Fraction Skills Inventory**

1. $\frac{5}{8}$ pound
2. $\frac{5}{6}$, $\frac{3}{4}$
3. $\frac{2}{3}$
4. $2\frac{2}{5}$
5. $\frac{38}{7}$
6. $\frac{12}{18}$
7. $8\frac{1}{5}$
8. $11\frac{11}{12}$
9. less than 1
10. $26\frac{13}{16}$ pounds
11. $6\frac{1}{4}$ hours
12. $6\frac{7}{20}$
13. $4\frac{1}{2}$
14. $3\frac{13}{21}$
15. $124\frac{1}{2}$ pounds
16. $27\frac{1}{2}$ inches
17. $\frac{21}{40}$
18. $\frac{1}{4}$
19. 15
20. $2 \times 3 = 6$
21. 28 crates
22. $37\frac{1}{2}$ hours
23. $1\frac{1}{9}$
24. 16
25. $\frac{1}{18}$
26. $2\frac{3}{16}$
27. 48
28. 28 bags
29. 10 pieces

**Page 11**

1. $\frac{1}{6}$
2. $\frac{5}{8}$
3. $\frac{3}{4}$
4. $\frac{3}{8}$
5. $\frac{2}{5}$
6. $\frac{4}{9}$
7. $\frac{5}{6}$
8. $\frac{2}{3}$
9. $\frac{5}{9}$

**Page 12**

1. $\frac{5}{12}$
2. $\frac{47}{100}$
3. $\frac{9}{16}$
4. $\frac{23}{36}$
5. $\frac{7}{12}$
6. $\frac{8}{25}$
7. $\frac{43}{60}$
8. $\frac{1,351}{2,000}$
9. $\frac{2}{5}$
10. $\frac{63}{100}$
11. $\frac{3}{4}$
12. $\frac{217}{720}$
13. $\frac{77}{280}$

**Page 13**

1. I P M I
2. P I I M
3. M P P I
4. P I M P

**Page 14**

1. $\frac{6}{12}$, $\frac{9}{18}$, $\frac{8}{16}$, $\frac{12}{24}$
2. $\frac{2}{7}$, $\frac{3}{8}$, $\frac{4}{9}$, $\frac{9}{20}$, $\frac{4}{15}$
3. $\frac{3}{4}$, $\frac{7}{8}$, $\frac{9}{15}$, $\frac{8}{12}$, $\frac{17}{30}$
4. $= \frac{1}{2}$ $\quad < \frac{1}{2}$ $\quad > \frac{1}{2}$

**Pages 16–17**

1. $\frac{1}{2}$ $\frac{1}{4}$ $\frac{1}{3}$ $\frac{1}{5}$ $\frac{1}{8}$
2. $\frac{5}{6}$ $\frac{8}{9}$ $\frac{8}{9}$ $\frac{7}{8}$ $\frac{2}{3}$
3. $\frac{2}{5}$ $\frac{1}{3}$ $\frac{7}{20}$ $\frac{9}{14}$ $\frac{8}{17}$

4. $\frac{3}{7}$ $\frac{3}{4}$ $\frac{1}{2}$ $\frac{3}{4}$ $\frac{3}{4}$
5. $\frac{15}{16}$ $\frac{7}{8}$ $\frac{4}{5}$ $\frac{2}{3}$ $\frac{9}{16}$
6. $\frac{1}{2}$ $\frac{1}{3}$ $\frac{1}{50}$ $\frac{7}{9}$ $\frac{5}{7}$
7. $\frac{2}{3}$
8. $\frac{3}{5}$
9. $\frac{2}{5}$
10. $\frac{1}{10}$
11. $\frac{5}{8}$
12. $\frac{2}{3}$
13. $\frac{1}{3}$
14. $\frac{1}{4}$
15. $\frac{3}{4}$
16. $\frac{1}{8}$
17. $\frac{5}{9}$
18. $\frac{4}{9}$
19. $\frac{5}{8}$
20. $\frac{3}{8}$

## Page 18

1. $\frac{24}{30}$ $\frac{18}{20}$ $\frac{3}{18}$ $\frac{20}{32}$
2. $\frac{20}{35}$ $\frac{18}{36}$ $\frac{14}{21}$ $\frac{54}{66}$
3. $\frac{25}{45}$ $\frac{33}{44}$ $\frac{35}{60}$ $\frac{15}{45}$
4. $\frac{25}{50}$ $\frac{15}{40}$ $\frac{16}{72}$ $\frac{35}{42}$

## Page 20

1. $1\frac{3}{4}$ $5\frac{1}{2}$ $2\frac{4}{5}$ $4\frac{2}{7}$ 4
2. $3\frac{1}{3}$ $3\frac{1}{4}$ 3 $3\frac{3}{5}$ 2
3. $1\frac{1}{12}$ 5 $7\frac{1}{2}$ $2\frac{2}{3}$ $4\frac{2}{3}$
4. 3 $2\frac{4}{5}$ $5\frac{1}{2}$ $2\frac{3}{4}$ $12\frac{1}{2}$
5. $1\frac{2}{3}$
6. $2\frac{2}{3}$
7. $1\frac{1}{4}$
8. $8\frac{1}{2}$

## Page 21

1. $\frac{11}{4}$ $\frac{11}{7}$ $\frac{16}{3}$ $\frac{44}{7}$ $\frac{23}{5}$
2. $\frac{19}{2}$ $\frac{61}{8}$ $\frac{29}{10}$ $\frac{35}{4}$ $\frac{32}{9}$
3. $\frac{31}{3}$ $\frac{57}{5}$ $\frac{53}{12}$ $\frac{55}{8}$ $\frac{49}{4}$
4. $\frac{19}{5}$ $\frac{43}{7}$ $\frac{75}{8}$ $\frac{25}{12}$ $\frac{79}{15}$

## Pages 22–24

1. $\frac{5}{9}$ $\frac{4}{7}$ $\frac{7}{8}$ $\frac{7}{12}$ $\frac{11}{15}$
2. $\frac{9}{16}$ $\frac{8}{9}$ $\frac{13}{15}$ $\frac{19}{20}$ $\frac{17}{24}$
3. $7\frac{3}{5}$ $14\frac{9}{10}$ $9\frac{9}{16}$ $14\frac{11}{15}$
4. $12\frac{8}{9}$ $17\frac{19}{24}$ $20\frac{6}{7}$ $15\frac{9}{10}$
5. $\frac{3}{4}$ $\frac{2}{3}$ $\frac{2}{3}$ $\frac{2}{3}$ $\frac{1}{2}$
6. $\frac{4}{5}$ $\frac{4}{5}$ $\frac{7}{10}$ $\frac{3}{4}$ $\frac{7}{9}$
7. $11\frac{1}{2}$ $16\frac{2}{3}$ $18\frac{6}{7}$ $42\frac{2}{3}$
8. $24\frac{2}{3}$ $17\frac{4}{5}$ $17\frac{3}{4}$ $31\frac{3}{4}$
9. $1\frac{2}{5}$ $1\frac{3}{8}$ $1\frac{3}{10}$ $1\frac{4}{9}$ 1
10. $1\frac{1}{3}$ $1\frac{1}{7}$ $1\frac{1}{2}$ $1\frac{1}{5}$ $1\frac{1}{2}$
11. $2\frac{2}{9}$ 2 $1\frac{1}{2}$ $2\frac{1}{4}$ $2\frac{1}{4}$
12. $17\frac{7}{8}$ $18\frac{2}{3}$ $18\frac{1}{10}$ 18

## Page 25

1. $1\frac{1}{4}$ $1\frac{1}{2}$ $1\frac{5}{8}$ $1\frac{1}{6}$ $1\frac{2}{9}$
2. $1\frac{5}{8}$ $1\frac{1}{3}$ $1\frac{4}{5}$ $1\frac{1}{3}$ $1\frac{1}{5}$
3. $1\frac{2}{5}$ 2 $\frac{11}{12}$ $1\frac{2}{3}$ $1\frac{2}{5}$

## Pages 26–28

1. $1\frac{4}{15}$ $1\frac{1}{12}$ $\frac{9}{10}$ $\frac{16}{21}$ $1\frac{7}{30}$
2. $\frac{6}{7}$ $1\frac{19}{30}$ 1 $1\frac{4}{15}$ $1\frac{11}{36}$
3. $1\frac{7}{12}$ $1\frac{5}{18}$ $1\frac{9}{20}$ $\frac{35}{36}$ $1\frac{5}{24}$
4. $2\frac{1}{24}$ $1\frac{11}{20}$ $2\frac{17}{36}$ $1\frac{7}{16}$ $1\frac{5}{9}$
5. $1\frac{1}{2}$ $1\frac{23}{24}$ $1\frac{3}{4}$ $1\frac{1}{40}$ $1\frac{1}{36}$
6. $1\frac{17}{18}$ $2\frac{3}{28}$ $\frac{13}{18}$ $1\frac{52}{63}$ $1\frac{43}{48}$
7. $11\frac{7}{20}$ $17\frac{7}{24}$ $12\frac{2}{9}$
8. $13\frac{19}{24}$ $18\frac{13}{21}$ $10\frac{11}{12}$
9. $16\frac{3}{20}$ $16\frac{5}{18}$ $13\frac{5}{24}$
10. $11\frac{1}{18}$ $15\frac{19}{40}$ $12\frac{17}{30}$
11. $17\frac{29}{30}$ $19\frac{7}{24}$ $17\frac{29}{36}$
12. $22\frac{1}{12}$ $13\frac{1}{24}$ $21\frac{19}{36}$
13. $15\frac{9}{16}$ $23\frac{1}{4}$ $15\frac{11}{12}$
14. $24\frac{11}{70}$ $24\frac{11}{18}$ $17\frac{45}{56}$

## Page 29

1. $< 1$ $= 1$ $> 1$ $= 1$
2. $< 1$ $> 1$ $> 1$ $< 1$
3. 5 9 12
4. 7 14 8
5. 6 17 13

Exact answers are given below.

1. $\frac{13}{16}$ 1 $1\frac{2}{5}$ 1
2. $\frac{7}{8}$ $1\frac{1}{2}$ $1\frac{17}{30}$ $\frac{59}{72}$
3. $4\frac{1}{4}$ $9\frac{11}{24}$ $11\frac{1}{2}$
4. $6\frac{4}{5}$ $14\frac{7}{40}$ $7\frac{2}{3}$
5. $6\frac{19}{30}$ $16\frac{3}{14}$ $13\frac{3}{20}$

## Page 30

1. $68\frac{1}{4}$ inches est. 69 inches
2. $16\frac{1}{10}$ miles est. 17 miles
3. $22\frac{7}{20}$ pounds est. 23 pounds
4. $8\frac{19}{24}$ pounds est. 9 pounds
5. $14\frac{1}{4}$ hours est. 15 hours
6. $128\frac{7}{10}$ pounds est. 129 pounds
7. $2\frac{23}{30}$ hours est. 4 hours

## Page 31

1. $\frac{1}{3}$ $\frac{1}{10}$ $\frac{1}{2}$ $\frac{1}{6}$ $\frac{3}{10}$
2. $\frac{1}{3}$ $\frac{3}{8}$ $\frac{1}{2}$ $\frac{1}{3}$ $\frac{1}{5}$
3. $3\frac{4}{7}$ 6 $1\frac{1}{3}$ $4\frac{2}{5}$
4. $8\frac{1}{4}$ $9\frac{1}{2}$ $9\frac{2}{5}$ $16\frac{2}{3}$

## Pages 32–33

1. $\frac{1}{4}$ $\frac{3}{8}$ $\frac{1}{2}$ $\frac{9}{16}$ $\frac{1}{5}$
2. $\frac{5}{12}$ $\frac{7}{15}$ $\frac{5}{24}$ $\frac{7}{30}$ $\frac{7}{18}$
3. $6\frac{13}{24}$ $6\frac{3}{14}$ $7\frac{26}{45}$ $3\frac{1}{20}$
4. $6\frac{11}{56}$ $15\frac{23}{36}$ $7\frac{13}{30}$ $12\frac{1}{4}$
5. $5\frac{1}{4}$ $3\frac{3}{8}$ $2\frac{1}{15}$
6. $2\frac{5}{16}$ $9\frac{1}{5}$ $5\frac{14}{45}$
7. $1\frac{1}{28}$ $7\frac{1}{30}$ $7\frac{7}{24}$
8. $4\frac{1}{12}$ $7\frac{11}{24}$ $7\frac{1}{18}$
9. $1\frac{23}{72}$ $4\frac{7}{12}$ $2\frac{11}{24}$
10. $6\frac{5}{24}$ $3\frac{1}{30}$ $2\frac{7}{20}$
11. $8\frac{1}{12}$ $8\frac{1}{30}$ $4\frac{1}{18}$
12. $2\frac{5}{24}$ $7\frac{2}{15}$ $9\frac{3}{10}$
13. $6\frac{1}{6}$ $4\frac{1}{10}$ $11\frac{5}{24}$
14. $3\frac{7}{30}$ $5\frac{1}{9}$ $3\frac{9}{40}$

## Pages 35–36

1. $7\frac{1}{6}$ $3\frac{4}{7}$ $11\frac{1}{2}$ $8\frac{3}{5}$ $9\frac{3}{11}$
2. $3\frac{4}{7}$ $3\frac{1}{3}$ $1\frac{5}{12}$ $3\frac{4}{9}$ $6\frac{11}{16}$
3. $3\frac{2}{3}$ $6\frac{1}{2}$ $7\frac{2}{3}$ $6\frac{2}{5}$
4. $4\frac{14}{15}$ $15\frac{5}{7}$ $6\frac{2}{3}$ $7\frac{3}{4}$
5. $10\frac{1}{2}$ $27\frac{2}{5}$ $\frac{7}{10}$ $2\frac{2}{3}$
6. $2\frac{5}{12}$ $11\frac{1}{2}$ $4\frac{7}{10}$ $2\frac{11}{12}$
7. $5\frac{13}{20}$ $4\frac{5}{8}$ $4\frac{7}{9}$
8. $14\frac{5}{12}$ $18\frac{13}{14}$ $8\frac{7}{12}$
9. $5\frac{4}{5}$ $31\frac{38}{45}$ $8\frac{1}{2}$
10. $7\frac{23}{40}$ $13\frac{19}{30}$ $1\frac{7}{12}$
11. $\frac{23}{30}$ $3\frac{25}{36}$ $13\frac{33}{40}$
12. $6\frac{17}{24}$ $10\frac{17}{30}$ $6\frac{7}{8}$
13. $13\frac{5}{6}$ $\frac{7}{8}$ $1\frac{37}{56}$
14. $16\frac{13}{20}$ $5\frac{7}{18}$ $14\frac{25}{48}$

## Page 37

1. c.
2. a.
3. b.
4. b.
5. a.
6. 3 3 5
7. 8 2 4
8. 1 6 4

Exact answers

1. $3\frac{5}{8}$
2. $1\frac{1}{12}$
3. $14\frac{1}{10}$
4. $1\frac{1}{4}$
5. $2\frac{31}{48}$
6. $2\frac{3}{4}$ $2\frac{5}{8}$ $4\frac{8}{9}$
7. $7\frac{7}{12}$ $2\frac{1}{6}$ $4\frac{1}{3}$
8. $\frac{11}{16}$ $6\frac{8}{15}$ $4\frac{4}{15}$

## Pages 38–39

1. $20\frac{7}{8}$ inches est. 21 inches
2. $154\frac{1}{4}$ pounds est. 154 pounds
3. $489\frac{2}{5}$ miles est. 490 miles
4. $3\frac{1}{3}$ yards est. 3 yards
5. $8\frac{1}{12}$ feet est. 8 feet
6. $3\frac{2}{3}$ rpm's est. 4 rpm's
7. $55\frac{3}{8}$ pounds est. 55 pounds

8. $1\frac{1}{8}$ pounds  est. 1 pound
9. $43\frac{1}{6}$ feet
10. $16\frac{5}{6}$ feet
11. $7\frac{3}{4}$ miles
12. $2\frac{1}{4}$ miles
13. \$$2\frac{3}{8}$ million
14. \$$\frac{5}{8}$ million
15. No. The total dimensions are $38\frac{1}{8}$ inches.
16. Yes. The total dimensions are $36\frac{1}{2}$ inches.

## Page 40

1. $\frac{8}{15}$  $\frac{10}{63}$  $\frac{7}{80}$  $\frac{15}{32}$
2. $\frac{1}{15}$  $\frac{16}{63}$  $\frac{25}{48}$  $\frac{9}{40}$
3. $\frac{14}{45}$  $\frac{21}{64}$  $\frac{5}{36}$  $\frac{16}{81}$
4. $\frac{9}{40}$  $\frac{5}{42}$  $\frac{10}{81}$
5. $\frac{16}{75}$  $\frac{14}{135}$  $\frac{8}{45}$

## Pages 41–42

1. $\frac{3}{10}$  $\frac{1}{6}$  $\frac{7}{16}$  $\frac{5}{14}$
2. $\frac{2}{15}$  $\frac{2}{9}$  $\frac{3}{8}$  $\frac{3}{4}$
3. $\frac{1}{6}$  $\frac{3}{8}$  $\frac{1}{6}$  $\frac{3}{4}$
4. $\frac{9}{20}$  $\frac{7}{80}$  $\frac{7}{10}$  $\frac{1}{6}$
5. $\frac{3}{10}$  $\frac{5}{28}$  $\frac{2}{5}$  $\frac{3}{20}$
6. $\frac{1}{10}$  $\frac{2}{7}$  $\frac{3}{4}$  $\frac{1}{16}$
7. $\frac{7}{32}$  $\frac{1}{9}$  $\frac{7}{15}$
8. $\frac{9}{40}$  $\frac{7}{60}$  $\frac{3}{32}$
9. $\frac{7}{45}$  $\frac{9}{100}$  $\frac{11}{72}$
10. $\frac{1}{8}$  $\frac{1}{9}$  $\frac{9}{32}$
11. $\frac{1}{12}$  $\frac{7}{20}$  $\frac{1}{15}$

## Page 43

1. $1\frac{5}{7}$  $2\frac{1}{4}$  $6\frac{2}{3}$  $2\frac{2}{5}$
2. 10  10  $1\frac{4}{5}$  24
3. 21  $5\frac{1}{2}$  14  $11\frac{1}{4}$
4. $8\frac{1}{6}$  $3\frac{1}{3}$  21  $1\frac{4}{5}$

## Page 44

1. $\frac{3}{8}$  $\frac{10}{21}$  $2\frac{3}{16}$  $1\frac{13}{20}$
2. $1\frac{2}{3}$  $\frac{3}{4}$  $4\frac{3}{8}$  $5\frac{1}{7}$
3. $2\frac{4}{5}$  25  $16\frac{1}{4}$  $38\frac{1}{2}$
4. 4  $22\frac{1}{2}$  84

## Page 45

1. $<$  $=$  $<$
2. $>$  $<$  $=$

Estimates and exact answers are given.

3. 18 and $16\frac{1}{2}$  12 and $12\frac{1}{2}$  45 and 48  16 and $14\frac{2}{3}$
4. 25 and 23  18 and $18\frac{2}{3}$  24 and $22\frac{1}{2}$  60 and $61\frac{1}{2}$
5. 4 and $4\frac{1}{6}$  10 and 12  15 and $12\frac{3}{5}$  4 and 5
6. 6 and 6  16 and $16\frac{2}{3}$  12 and $9\frac{5}{8}$  9 and $8\frac{5}{8}$

## Pages 46–47

1. 171 inches  est. 174 inches
2. 32 hours  *part* of
3. 184 miles  *part* of
4. \$745  *part* of
5. $78\frac{1}{8}$ pounds  est. 63 pounds
6. 27 miles  est. 36 miles
7. \$27  *part* of
8. \$11.50  est. \$12
9. \$52
10. $9\frac{1}{2}$ yards
11. $2\frac{3}{4}$ yards
12. \$67.20
13. \$12.60
14. \$44.10
15. \$111.30
16. \$18,800
17. \$9,800
18. 280 people
19. 105 people
20. 35 people

## Page 49

1. $1\frac{1}{14}$  $\frac{2}{3}$  $\frac{5}{9}$  $\frac{7}{12}$
2. $1\frac{1}{10}$  4  $1\frac{1}{20}$  $1\frac{1}{5}$
3. $2\frac{1}{10}$  $2\frac{1}{3}$  $1\frac{2}{7}$  $1\frac{1}{5}$
4. $\frac{3}{28}$  4  $1\frac{2}{5}$  $2\frac{2}{15}$
5. $\frac{5}{9}$  $\frac{13}{40}$  $8\frac{1}{2}$  $1\frac{1}{6}$
6. $\frac{7}{20}$  $2\frac{2}{9}$  $\frac{24}{25}$  $\frac{3}{5}$

## Page 51

1. 30  27  $6\frac{2}{3}$  9
2. $21\frac{1}{3}$  30  34  36
3. 50  84  $37\frac{1}{2}$  120
4. 90  36  $26\frac{2}{3}$  32
5. 60  300  48  $13\frac{1}{3}$
6. 32  $5\frac{1}{3}$  $16\frac{4}{5}$  $16\frac{2}{3}$

## Page 53

1. $\frac{1}{5}$ $\frac{1}{9}$ $\frac{1}{48}$ $\frac{1}{8}$
2. $\frac{1}{36}$ $\frac{3}{28}$ $\frac{1}{12}$ $\frac{3}{10}$
3. $\frac{1}{22}$ $\frac{1}{25}$ $\frac{1}{8}$ $\frac{2}{75}$
4. $\frac{1}{32}$ $\frac{1}{40}$ $\frac{2}{75}$ $\frac{3}{125}$
5. $\frac{1}{15}$ $\frac{3}{20}$ $\frac{5}{64}$ $\frac{1}{32}$
6. $\frac{1}{99}$ $\frac{3}{64}$ $\frac{1}{54}$ $\frac{3}{80}$

## Pages 54–55

1. 2 $2\frac{1}{2}$ $4\frac{2}{5}$ $19\frac{1}{2}$
2. $\frac{2}{5}$ $\frac{5}{6}$ $\frac{4}{21}$ $\frac{4}{27}$
3. $\frac{1}{2}$ $\frac{4}{5}$ $\frac{7}{30}$ $\frac{3}{20}$
4. $7\frac{1}{2}$ $8\frac{3}{4}$ $4\frac{4}{5}$ $6\frac{2}{5}$
5. $3\frac{1}{3}$ $3\frac{6}{7}$ $1\frac{7}{15}$ $\frac{4}{5}$
6. $1\frac{1}{8}$ 2 $1\frac{3}{5}$ $2\frac{4}{5}$
7. 3 $\frac{1}{2}$ $1\frac{19}{21}$ 4
8. $2\frac{2}{3}$ 3 $1\frac{5}{28}$ $\frac{1}{3}$
9. $2\frac{11}{12}$ $1\frac{13}{42}$ $5\frac{7}{18}$ 6
10. $3\frac{1}{5}$ $2\frac{2}{3}$ $2\frac{4}{5}$ $2\frac{13}{36}$

## Page 56

1. 15 88
2. 27 72
3. 20 64
4. 75 24
5. 96 66
6. 100 36
7. 150 60

## Page 57

1. $<$ $>$ $>$
2. $>$ $<$ $>$

Exact answers and estimates are given.

3. $3\frac{1}{2}$ and 4 12 and $12\frac{1}{2}$ $\frac{2}{3}$ and $\frac{3}{4}$
4. $2\frac{2}{3}$ and $2\frac{1}{2}$ $2\frac{1}{2}$ and $2\frac{3}{4}$ $2\frac{1}{4}$ and $2\frac{1}{3}$
5. $1\frac{1}{2}$ and $1\frac{1}{3}$ $\frac{5}{12}$ and $\frac{2}{5}$ $1\frac{13}{20}$ and $1\frac{3}{4}$
6. $\frac{8}{15}$ and $\frac{1}{2}$ 5 and $5\frac{5}{7}$ $\frac{7}{10}$ and $\frac{3}{4}$

## Pages 58–59

1. 6 pieces est. $5\frac{5}{8}$
2. $1\frac{1}{2}$ pounds est. $1\frac{2}{3}$
3. 6 suits est. $5\frac{1}{2}$
4. 6 loaves est. $4\frac{1}{2}$
5. 32 cans est. 24
6. 9 strips $9\frac{3}{8}$
7. 4 dresses (with some material left) est. $5\frac{1}{2}$
8. $1\frac{5}{6}$ pounds est. 2
9. 10 repairs
10. 50 repairs
11. 27 students
12. 9 students
13. 150 workers
14. 60 workers
15. $5\frac{1}{4}$ pounds
16. $1\frac{3}{4}$ pounds
17. \$$\frac{3}{4}$ million
18. \$$\frac{1}{2}$ million

## Pages 60–61, Fractions Review

1. $\frac{7}{12}$
2. $\frac{7}{15}$, $\frac{4}{9}$
3. $\frac{9}{16}$
4. $4\frac{1}{6}$
5. $\frac{93}{8}$
6. $\frac{9}{12}$
7. $19\frac{7}{8}$
8. $20\frac{23}{30}$
9. greater than 1
10. $2\frac{5}{12}$ hours
11. $13\frac{5}{16}$ pounds
12. $5\frac{19}{40}$
13. $5\frac{4}{9}$
14. $8\frac{7}{18}$
15. $\frac{3}{4}$ pound
16. $1\frac{5}{6}$ miles
17. $\frac{8}{15}$
18. $\frac{7}{40}$
19. $3\frac{1}{2}$
20. 18
21. 65 pounds
22. $7\frac{7}{8}$ miles
23. $\frac{32}{35}$
24. $31\frac{1}{2}$
25. $\frac{3}{20}$
26. $\frac{15}{28}$
27. 75
28. 6 bookcases
29. \$16

## Pages 62–63, Decimal Skills Inventory

1. 0.076
2. $3\frac{1}{25}$
3. $0.41\frac{2}{3}$
4. 0.08
5. 0.07, 0.08, 0.087, 0.7
6. 2.1
7. 4.439
8. 23.953
9. 12.1 pounds
10. 84.1°
11. 9.143
12. 11.63
13. 1.15 yards
14. 286.8 miles

15. 10.971

16. 0.0228

17. 49.2

18. 144 miles

19. $ 91.35

20. 0.86

21. 7.3

22. 370

23. 200

24. 0.496

25. $2.60

26. 5 yards

**Page 65**

1. 3 5 1 6
2. 4 2 3 1
3. 6 7 3 5
4. 9 7 2 3
5. 4 9 1 3
6. 3 8 5 6
7. 9 tenths
8. 7 thousandths

**Page 66**

1. tenths tenths hundredths
2. hundredths hundredths thousandths
3. thousandths ten-thousandths
4. thousandths ten-thousandths
5. hundredths thousandths
6. tenths ten-thousandths
7. five tenths two and nine tenths
8. seven hundredths
   three and twelve hundredths
9. sixteen thousandths
   one and three thousandths
10. nine thousandths
    seven and twenty-one thousandths
11. sixteen ten-thousandths
    ten and four hundred two thousandths

**Page 67**

1. 0.7 4.9
2. 0.63 0.022
3. 0.02 0.85
4. 0.4 0.019
5. 12.3 0.41
6. 1.05 0.206
7. 48.9 0.008
8. 0.304 11.07
9. 0.0015 4.013
10. 230.1 0.0071
11. 0.047 8.0002
12. 0.00005 0.000019

**Page 68**

1. **c.** 5.006
2. **a.** 3.105
3. **d.** 700.4
4. **b.** 40.092
5. **d.** both a and b
6. **b.** 0.034

**Page 69**

1. $\frac{2}{25}$ $\frac{3}{8}$ $\frac{3}{625}$
2. $3\frac{3}{5}$ $9\frac{43}{50}$ $10\frac{1}{500}$
3. $\frac{17}{200}$ $5\frac{2}{25}$ $\frac{1}{400}$
4. $7\frac{1}{5}$ $\frac{3}{20}$ $8\frac{4}{25}$
5. $\frac{81}{25{,}000}$ $19\frac{393}{5{,}000}$ $123\frac{231}{500}$
6. $16\frac{1}{25{,}000}$ $7\frac{11}{50}$ $3\frac{1}{125{,}000}$
7. $2{,}036\frac{4}{5}$ $48\frac{1}{50}$ $3\frac{3}{40}$

**Page 70**

1. 0.25 0.4 0.625 or $0.62\frac{1}{2}$ 3.5
2. $0.22\frac{2}{9}$ 0.24 $0.16\frac{2}{3}$ 1.375 or $1.37\frac{1}{2}$
3. $0.83\frac{1}{3}$ 0.3 $0.57\frac{1}{7}$ $1.83\frac{1}{3}$
4. 0.6 0.9 $0.08\frac{1}{3}$ 2.25
5. $0.13\frac{1}{3}$ 0.75 0.125 or $0.12\frac{1}{2}$ 1.3

## Page 71

1. 0.04 0.99 0.707
2. 0.33 0.11 0.2
3. 0.006 0.4 0.07
4. 0.03, 0.033, 0.303, 0.33 0.08, 0.082, 0.28, 0.8
5. 0.061, 0.106, 0.16, 0.6 0.007, 0.017, 0.02, 0.2
6. 0.045, 0.4, 0.405, 0.45 0.04, 0.304, 0.32, 0.4
7. 0.0072, 0.02, 0.027, 0.07 0.026, 0.06, 0.0602, 0.2

## Page 73

1. 4.3 0.6 516.2 27.1
2. 0.58 12.49 0.09 2.10
3. 23.486 0.054 5.407 0.003
4. 17 2 357 200
5. $14 $10 $2 $347
6. $2.80 $63.50 $1.00 $5.30
7. $0.99 $4.68 $23.95 $1.09
8. 49.1 49.08 49.076 49.0758 49
9. 0.63 0.67 0.43 0.78
10. 0.083 0.833 0.188 0.133

## Page 74

1. 1.207 25.18
2. 134.64 39.617
3. 24.313 65.303
4. 75.449 18.8097
5. 132.543 93.277
6. 23.458 0.5069
7. 37.815 22.135

## Page 75

1. 201.4
2. 201
3. 170
4. 201.398
5. 4.21
6. 4.2
7. 4
8. 4.207
9. 19.2
10. 19
11. 18
12. 19.237

## Page 76

1. 10.66 inches est. 11 inches
2. 1,200 miles est. 1,200 miles
3. 76.9 inches est. 80 inches
4. 9.7 pounds est. 10 pounds
5. 121.5 miles est. 120 miles
6. 103.1° est. 105°
7. 8.02 kilograms est. 7 kilograms
8. 153.1 million est. 140 million

## Page 77

1. 0.44 0.6367 0.52
2. 0.122 54.36 88.534
3. 11.064 0.953 12.068
4. 7.769 0.068 2.507
5. 88.515 3.0014 0.0708
6. 0.236 0.017 1.193
7. 0.0641 0.422 0.541

## Page 78

1. 8.6
2. 8
3. 8.1
4. 8.544
5. 31.62
6. 31.6
7. 32
8. 30
9. 31.628
10. 158.2
11. 159
12. 160
13. 140
14. 158.268

## Pages 79–80

1. 5.1 years; est. 5 years
2. 79.5 sq mi; est. 79 sq mi
3. 1.33 meters; est. 1 meter
4. 12.8 miles; est. 13 miles
5. 402.7 miles; est. 402 miles
6. 123.15 meters; est. 123 meters
7. 3.3 million; est. 3 million
8. $0.35; est. $0.30

9. 1.89 kilograms
10. 0.11 kilogram under
11. $0.32
12. $0.81
13. 88.5 million
14. 6.9 inches
15. 6.5 inches
16. 1.05 million

## Pages 81–83

1. 15.2 8.28 40.2 42.4 5.88
2. 1.23 39 5.31 34.4 1.14
3. 277.6 20.23 3.306 542.7 28.72
4. 63.42 1.008 626.4 39.12 1.401
5. 142.5 27.72 1,341 560.7 550.2
6. 0.054 0.035 0.0012 0.0032 0.0016
7. 1.12 0.0584 6.44 0.0348 0.198
8. 29.26 0.312 0.258 4.428 1.388
9. 11.7 2.496 8.188 2.414 0.2184
10. 0.486 2.184 26.46 0.1411 6.072
11. 12.04 4.592 4.104
12. 0.5589 1.692 8.235
13. 48.84 296.64 0.46774
14. 66.138 253.176 273.604
15. 57.932 3.66912 7,252.3
16. 36.2142 72.89 3.44787
17. 2.7712 0.14904 44.7468
18. 836.368 0.047413 112.274
19. 8.29376 17,210.5704 2.8764

## Page 84

1. 8 0.9 36.4 7.21
2. 3 27.5 890 86.3
3. 900 2,360 475 1,600
4. 3.4 124 3,850 60
5. $12.50 $60 $2,250 $30

## Page 85

Estimates and exact answers are given.

1. $>$ $<$ $=$
2. $<$ $>$ $>$
3. 1.5 and 1.41 0.072 and 0.0747 4.8 and 5.12
4. 0.48 and 0.474 0.18 and 0.174 1.4 and 1.274
5. 20 and 19.24 2.4 and 2.542 0.35 and 0.3922
6. 0.28 and 0.2812 6.3 and 6.097 0.32 and 0.3564
7. 10 and 9.9164 16 and 17.507 6.3 and 6.2652

## Pages 86–87

1. $144 est. $160
2. 1,061.5 miles est. 1,200 miles
3. 22,540 est. 20,000
4. $6.72 est. $7.20
5. 517.5 pounds est. 500 pounds
6. 13.5 liters est. 16 liters
7. $9.66 est. $8
8. $403.13 est. $400
9. 3,625 pounds est. 3,600 pounds
10. 3.3 gallons
11. 9.9 gallons
12. 83.2 calls
13. 1.5 hours
14. 72 kilograms
15. $135.10
16. $151.03
17. 1,560 kilograms
18. 90 kilograms
19. 198 pounds
20. No, he's too heavy.

## Page 88

1. 2.3 7.8 0.096 8.25
2. 0.024 49.6 3.89 0.631
3. 4.8 0.36 0.08 5.7
4. 0.183 36.4 6.03 0.517

### Pages 89–90

**1.** 9.6  38  0.7  6.9

**2.** 0.28  14  0.83  0.09

**3.** 14.7  1.93  605  12.8

**4.** 1.3  0.24  37  2.9

**5.** 760  480  2,700  9,640

**6.** 13,300  4  720  85

**7.** 325  71,400  935  2,080

**8.** 490  3,070  560  9,310

**9.** 0.06  2.07  40.4  0.182

### Page 91

**1.** 70  400  3,700  6,400

**2.** 9,000  8,000  3,900  6,280

**3.** 200  4,000  2,000  28

**4.** 460  7,200  500  420

**5.** 62,000  78,000  5,900  83,000

### Page 92

**1.** 0.09  3.6  2.73  0.004

**2.** 0.142  0.013  7.28  0.006

**3.** 0.0375  0.0018  0.002  0.428

**4.** 1.345  0.0032  6.954  0.0158

**5.** $0.29  $6.50  $0.54  $0.02

### Page 93

**1.** 1.1  11.4  3.3

**2.** 0.1  0.3  7.1

**3.** 3.1  0.2  138.5

**4.** 1.4  29.4  22.2

**5.** 1.8  1.4  78.8

### Page 94

**1.** $<$  $>$  $>$

**2.** $>$  $<$  $>$

Exact answers and estimates are given.

**3.** 4.6 and 5  0.15 and 0.16  6.2 and 4.5

**4.** 0.2 and 0.2  3.4 and 3.75  0.75 and 0.75

**5.** 0.24 and 0.2  1.5 and 1.25  16.5 and 15

**6.** 14 and 15  8 and 10  0.25 and 0.4

### Pages 95–96

**1.** 127.2 pounds  est. 125

**2.** 249.5 miles  est. 240

**3.** $9.60  est. $7.50

**4.** 38.5 hours  est. 50

**5.** 8.3 acres  est. 8

**6.** $1.60  est. $2

**7.** 23 miles

**8.** 18 inches

**9.** 0.327

**10.** 0.311

**11.** 35.4 points per game

**12.** 20.9 points per game

**13.** 20 months

**14.** 39 chairs per worker

**15.** 68 pages

### Pages 97–98, Cumulative Review

**1.** $8\frac{19}{24}$

**2.** $3\frac{17}{18}$

**3.** $13\frac{1}{3}$

**4.** 4

**5.** 0.0014

**6.** $\frac{2}{125}$

**7.** 0.4375

**8.** 0.02

**9.** 0.013, 0.03, 0.031, 0.1

**10.** 12.4

**11.** 21.356

**12.** 10.6973

**13.** 11.35 pounds

**14.** 12.203

**15.** 2.346

**16.** 1.984 inches

**17.** 2.844

**18.** 2,080

**19.** 720

**20.** 16.5 centimeters

**21.** $6.75

**22.** 4.09

**23.** 7.2

**24.** 6

**25.** 250

**26.** 0.00513

**27.** 34.5 mph

**28.** $15.70

### Pages 99–100, Percent Skills Inventory

**1.** 1.7%

**2.** 0.04

**3.** $62\frac{1}{2}\%$

**4.** $\frac{3}{25}$

**5.** $\frac{1}{12}$

**6.** 12

**7.** 11.5

**8.** 36

**9.** $0.84

**10.** $1,126

**11.** $13.25

**12.** $33\frac{1}{3}\%$

**13.** 8%

**14.** 11%

**15.** $66\frac{2}{3}\%$

**16.** 40%

**17.** 80

**18.** 56

**19.** $5,080

**20.** $26.50

**21.** 1,460 people

**Page 101**

1. 100
2. 49
3. 100
4. 100
5. 100
6. 50

**Page 102**

1. 32%   9%   60%   13.6%
2. 0.5%   $37\frac{1}{2}\%$   $8\frac{1}{3}\%$   4.5%
3. 0.16%   0.03%   2.5%   $33\frac{1}{3}\%$
4. 12.5%   3.75%   0.9%   20%
5. 0.1 = 10%   0.2 = 20%   0.01 = 1%
   0.3 = 30%   0.4 = 40%   0.25 = 25%
   0.7 = 70%   0.6 = 60%   0.5 = 50%
   0.9 = 90%   0.8 = 80%   0.75 = 75%

**Page 103**

1. 0.2   0.35   0.08   0.6
2. 0.035   0.004   0.0003   0.216
3. $0.62\frac{1}{2}$   $0.06\frac{2}{3}$   0.028   0.19
4. 0.07   0.015   2.0   0.142
5. 50% = 0.5   5% = 0.05   37.5% = 0.375
   25% = 0.25   1% = 0.01   62.5% = 0.625
   75% = 0.75   100% = 1.0   87.5% = 0.875
   20% = 0.2   80% = 0.8   12.5% = 0.125

**Page 104**

1. 40%   25%   $33\frac{1}{3}\%$   $37\frac{1}{2}\%$
2. 24%   $66\frac{2}{3}\%$   $83\frac{1}{3}\%$   $12\frac{1}{2}\%$
3. 90%   $87\frac{1}{2}\%$   55%   $41\frac{2}{3}\%$
4. $16\frac{2}{3}\%$   80%   70%   $8\frac{1}{3}\%$
5. $62\frac{1}{2}\%$   $44\frac{4}{9}\%$   $42\frac{6}{7}\%$   45%
6. 16%   30%   60%   18%

**Page 105**

1. $\frac{7}{20}$   $\frac{1}{5}$   $\frac{1}{8}$   $\frac{3}{50}$
2. $\frac{1}{6}$   $\frac{1}{100}$   $\frac{9}{10}$   $\frac{3}{8}$
3. $\frac{3}{25}$   $\frac{99}{100}$   $\frac{2}{3}$   $\frac{9}{200}$
4. $\frac{4}{5}$   $\frac{1}{3}$   $\frac{1}{25}$   $\frac{1}{12}$

**Page 106**

$\frac{1}{2} = 0.5 = 50\%$
$\frac{1}{4} = 0.25 = 25\%$
$\frac{3}{4} = 0.75 = 75\%$

$\frac{1}{5} = 0.2 = 20\%$
$\frac{2}{5} = 0.4 = 40\%$
$\frac{3}{5} = 0.6 = 60\%$
$\frac{4}{5} = 0.8 = 80\%$
$\frac{1}{3} = 0.33\frac{1}{3} = 33\frac{1}{3}\%$
$\frac{2}{3} = 0.66\frac{2}{3} = 66\frac{2}{3}\%$

$\frac{1}{8} = 0.125 \text{ or } 0.12\frac{1}{2} = 12\frac{1}{2}\%$
$\frac{3}{8} = 0.375 \text{ or } 0.37\frac{1}{2} = 37\frac{1}{2}\%$
$\frac{5}{8} = 0.625 \text{ or } 0.62\frac{1}{2} = 62\frac{1}{2}\%$
$\frac{7}{8} = 0.875 \text{ or } 0.87\frac{1}{2} = 87\frac{1}{2}\%$
$\frac{1}{10} = 0.1 = 10\%$
$\frac{3}{10} = 0.3 = 30\%$
$\frac{7}{10} = 0.7 = 70\%$
$\frac{9}{10} = 0.9 = 90\%$
$\frac{1}{6} = 0.16\frac{2}{3} = 16\frac{2}{3}\%$
$\frac{5}{6} = 0.83\frac{1}{3} = 83\frac{1}{3}\%$

**Pages 107–108**

1. 6   67.55   78
2. 7.2   7.5   24
3. 510   400   209
4. 168   403   799
5. 10.14   0.448   13.554
6. 25   12   14
7. 270   525   3
8. 21   60   175
9. 3   $3\frac{3}{4}$   110

**Page 109**

1. 24   230   7
2. 90   130   40
3. 90   400   30
4. 17   70   300
5. 120   241   900

### Page 110

1. 24 days
2. $4,750
3. 6 e-mails
4. 351 members
5. 6 classes
6. $294
7. $600
8. 10 days
9. $9\frac{3}{4}$ pounds

### Page 111

1. < > <
2. > = <
3. **c.** exact $22.43 est. $23
4. **c.** exact 3.36 est. 3.5
5. **b.** exact 396.32 est. 400
6. **a.** exact 267.12 est. 270
7. **c.** exact 1,020 est. 1,000
8. **b.** exact. $24.70 est. $26

### Page 112

1. $38.40 est. $40
2. $311.60 est. $304
3. $622.08 est. $638
4. $143,960 est. $144,000
5. 241,615 est. 237,500
6. $611.60 est. $630
7. $67.92 est. $68
8. 70 est. 64
9. 136 est. 136

### Pages 113–114

1. 50% 25% $33\frac{1}{3}$%
2. 10% $16\frac{2}{3}$% $12\frac{1}{2}$%
3. 80% 40% $66\frac{2}{3}$%
4. $83\frac{1}{3}$% 75% $87\frac{1}{2}$%
5. $87\frac{1}{2}$% 90% 80%
6. 7% 6% 9%
7. 35% 38% 32%
8. $58\frac{1}{3}$% 100% $44\frac{4}{9}$%
9. $83\frac{1}{3}$% $42\frac{6}{7}$% 90%
10. 2% 6% $8\frac{1}{3}$%
11. 20% 80% $22\frac{1}{2}$%
12. $16\frac{2}{3}$% 26% 0.396%

### Pages 115–116

1. 75%
2. 4%
3. $62\frac{1}{2}$%
4. 90%
5. 8%
6. 5%
7. $33\frac{1}{3}$%
8. $66\frac{2}{3}$%
9. 25%
10. 55%
11. 45%
12. $62\frac{1}{2}$%
13. $6\frac{1}{4}$%
14. 25%
15. 75%
16. $12\frac{1}{2}$%
17. 5%
18. $16\frac{2}{3}$%
19. $8\frac{1}{3}$%
20. $33\frac{1}{3}$%

### Pages 117–118

1. **a.** $95
   **b.** $380
   **c.** $\frac{1}{4}$
   **d.** 25%
2. **a.** 24
   **b.** 80
   **c.** $\frac{3}{10}$
   **d.** 30%
3. 21%
4. 30%
5. $12\frac{1}{2}$%
6. $33\frac{1}{3}$%
7. 125%
8. $44\frac{4}{9}$%
9. $12\frac{1}{2}$%
10. 15%

### Pages 119–120

1. 32 90
2. 64 125
3. 150 80
4. 63 24
5. 75 1,280
6. 1,800 1,400
7. 64 180
8. 480 240
9. 192 1,500
10. 305 1,050
11. 438 875

## Pages 121–122

1. $190
2. 30 problems
3. $7,200
4. 390 seats
5. $60
6. 120 members
7. $12,400
8. $170,000
9. 2,250 radios
10. 18,000 people
11. 3,000 miles
12. $2,000
13. $35
14. $910
15. $200.20

## Pages 123–124

1. %, 50%   *P*, 33
2. *P*, 90   %, 20%
3. *W*, 60   *P*, 340
4. *P*, 15   %, 80%
5. %, $33\frac{1}{3}$%   *W*, 460
6. *P*, 30 employees
7. %, 75%
8. *P*, $1.45
9. *W*, $3,200
10. %, 20%
11. *P*, 198 people
12. *W*, 1,200 seats

## Pages 125–126, Cumulative Review

1. $13\frac{7}{8}$
2. $4\frac{2}{5}$
3. $1\frac{3}{4}$
4. 1.30
5. 0.325
6. 1.5
7. 12.3%
8. 0.056
9. $58\frac{1}{3}$%
10. $\frac{21}{25}$
11. $\frac{5}{16}$
12. 24.3
13. 51.98
14. **c.** $72 \div 8$
15. $140.25
16. 15,000 voters
17. $62\frac{1}{2}$%
18. $8\frac{1}{2}$%
19. 75%
20. $62\frac{1}{2}$%
21. 200
22. 384
23. 120 miles
24. 320 jackets

## Pages 127–130, Posttest A

1. $\frac{4}{9}$
2. $20\frac{1}{24}$
3. less than 1
4. $59\frac{1}{6}$ inches
5. $42\frac{1}{6}$
6. $6\frac{15}{16}$ inches
7. $\frac{1}{15}$
8. $40\frac{1}{2}$
9. **b.** $4 \times 6 = 24$
10. $50.75
11. $\frac{1}{25}$
12. $1\frac{11}{45}$
13. **c.** more than 36 bags
14. $540
15. 50.208
16. 0.02, 0.021, 0.12, 0.2
17. 309.799
18. 128.1 pounds
19. 39.613
20. 17.8
21. 136.2 miles
22. 3.8912
23. 5.6
24. $3.60
25. 0.26
26. 19.3
27. 14.1
28. 15 miles
29. 0.079
30. $56\frac{1}{4}$%
31. $\frac{1}{6}$
32. 54.72
33. 12
34. **b.** $84 \div 4$
35. $221.44
36. **a.** 80% of $12,000
37. $83\frac{1}{3}$%
38. $62\frac{1}{2}$%
39. 160
40. $250,000

## Page 134

1. C—$1\frac{5}{8}$ inches
   D—2 inches
   E—$2\frac{1}{4}$ inches
   F—$2\frac{7}{8}$ inches
   G—$3\frac{1}{8}$ inches
   H—$3\frac{7}{16}$ inches
   I—$3\frac{15}{16}$ inches
   J—$4\frac{1}{2}$ inches
   K—$4\frac{13}{16}$ inches
   L—$5\frac{9}{16}$ inches
2. $\frac{3}{8}$ inch
3. $\frac{9}{16}$ inch
4. $1\frac{1}{16}$ inches
5. $3\frac{7}{16}$ inches
6. $2\frac{11}{16}$ inches

## Page 135

1. C—2.9 centimeters
   D—3.4 centimeters
   E—4.0 centimeters
   F—5.5 centimeters
   G—7.1 centimeters
   H—7.5 centimeters
   I—10.4 centimeters
   J—11.7 centimeters
   K—13.2 centimeters
   L—15.0 centimeters

2. 1.1 centimeters
3. 1.6 centimeters
4. 1.5 centimeters
5. 9.8 centimeters
6. 4.2 centimeters

## Pages 136–137

1. $21\frac{1}{3}$ inches
2. $18\frac{1}{4}$ feet
3. 39 inches
4. 83 feet
5. 24 meters
6. $228
7. $185\frac{1}{2}$ yards
8. 29.8 centimeters
9. 25 yards
10. 10 inches
11. 17 meters
12. 2.6 meters

## Pages 138–139

1. 20 sq ft
2. 17.36 sq cm
3. 30 sq in.
4. 28 sq ft
5. 39.6 sq m
6. $649.44
7. $7\frac{1}{2}$ sq yd
8. $186.75
9. 12.96 sq m
10. 0.16 sq m
11. 18 sq ft
12. $42\frac{1}{4}$ sq in.

## Pages 140–141

1. 10 cu ft
2. 531.25 cu ft
3. 918 cu ft
4. 594 cu in.
5. 221 cu in.
6. 1,500 cu in.
7. $7\frac{1}{2}$ cu in.
8. 200 boxes
9. 1,279.08 cu m
10. 84 cu ft
11. **c.** about 3 cu yd
12. **d.** 2,000

## Pages 142–143

1. 176 yards
2. 110 feet
3. $236.50
4. 62.8 feet
5. 31.4 inches
6. 11 feet
7. $12.98
8. 88 inches
9. 44 inches
10. 94.2 feet
11. 3.14 feet
12. $47

## Pages 144–145

1. 154 sq ft
2. 314 sq ft
3. 616 tiles
4. $147.84
5. 1,256 sq ft
6. 50 sq ft
7. $38\frac{1}{2}$ sq yd
8. $1,732.50
9. $\frac{11}{14}$ sq ft
10. 960 sq in.
11. 154 sq in.
12. 806 sq in.

## Page 146

1. 2 cups flour, 3 cups sugar, $\frac{1}{2}$ teaspoon salt, $2\frac{2}{3}$ cups egg whites, $3\frac{1}{3}$ teaspoons cream of tartar, $2\frac{1}{2}$ teaspoons vanilla
2. $\frac{1}{2}$ cup flour, $\frac{3}{4}$ cup sugar, $\frac{1}{8}$ teaspoon salt, $\frac{2}{3}$ cup egg whites, $\frac{5}{6}$ teaspoon cream of tartar, $\frac{5}{8}$ teaspoon vanilla
3. 5 cups flour, $7\frac{1}{2}$ cups sugar, $1\frac{1}{4}$ teaspoons salt, $6\frac{2}{3}$ cups egg whites, $8\frac{1}{3}$ teaspoons cream of tartar, $6\frac{1}{4}$ teaspoons vanilla
4. $1\frac{1}{2}$ cups flour, $2\frac{1}{4}$ cups sugar, $\frac{3}{8}$ teaspoon salt, 2 cups egg whites, $2\frac{1}{2}$ teaspoons cream of tartar, $1\frac{7}{8}$ teaspoons vanilla
5. 3

## Pages 148–149

1. 42 yards
2. $83\frac{1}{3}$ sq yd
3. 870 km
4. 38 cm
5. 150 cm
6. $\frac{1}{4}$ acre
7. 5,542 km
8. 27.9 cm; 21.6 cm
9. $\frac{1}{2}$ mile
10. 114 inches
11. 193 cm
12. $\frac{3}{4}$ pound
13. 135 cubic feet
14. $1\frac{3}{4}$ tons
15. 26 quarts
16. 7.7 pounds
17. $1\frac{1}{2}$ quarts
18. 4,200 grams

## Page 151

1. $80
2. $67.50
3. $5
4. $335
5. $81
6. $172.50
7. $79.80
8. $57

## Page 152

1. $18
2. $12.50
3. $105
4. $3.60
5. $18.40
6. $1,218.40

## Page 153

1. $162
2. $275
3. $17.50
4. $525
5. $2,525
6. $90

## Page 155

1. a. $12 b. $812
2. a. $12.18 b. $824.18
3. a. $12.36 b. $836.54
4. a. $12.55 b. $849.09
5. a. $24 b. $1,224
6. a. $24.48 b. $1,248.48
7. a. $24.97 b. $1,273.45
8. a. $25.47 b. $1,298.92
9. a. $1,000 b. $21,000
10. a. $1,050 b. $22,050
11. a. $1,102.50 b. $23,152.50
12. a. $1,157.63 b. $24,310.13

## Page 157

1. 25%
2. 50%
3. $33\frac{1}{3}$%
4. 40%
5. 25%
6. $33\frac{1}{3}$%
7. 35%
8. $16\frac{2}{3}$%
9. 28%
10. 20%
11. $97.98
12. $168.70

## Page 158

1. $19.50
2. $58.50
3. $16.65
4. $33.30
5. $150.30
6. $162.32
7. $187.32

## Page 159

1. 2.4 children/family
2. $21.74/hour
3. 26.5 students/classroom
4. 14.7 rooms/painter
5. 35.7 tickets/member
6. 372.2 miles/hour
7. 21.6 miles/gallon
8. 167 pieces/hour

## Page 161

1. a. $1.14$\frac{1}{2}$
   b. $ 1.03
   c. $0.99$\frac{1}{2}$
   d. Sam's Store
2. a. $0.35$\frac{4}{7}$
   b. $0.38$\frac{1}{6}$
   c. $0.32$\frac{1}{4}$
   d. Sam's Store
3. a. $1.29$\frac{1}{2}$
   b. $1.24$\frac{1}{2}$
   c. $1.31$\frac{2}{3}$
   d. Gert's Groceries
4. a. $1.49$\frac{1}{2}$
   b. $1.50
   c. $1.95
   d. Fred's and Gert's are the same.

## Page 163

1. $2,017.25
2. $10,908.75
3. $591
4. $17,677.25
5. $43,196.75
6. $4,052.50
7. $6,680
8. $25,450
9. $11,922.50
10. $3,046

**Pages 164–165**

1. \$35,000 \$2,170 \$508
2. \$29,300 \$1,817 \$425
3. \$38,460 \$2,385 \$558
4. \$46,080 \$2,857 \$668
5. \$59,900 \$3,714 \$869
6. \$32,850 \$2,037 \$476

**Pages 166–167**

1. \$14,000
2. \$36,000
3. Rent \$7,200
   Food \$11,160
   Utilities \$1,800
   Entertainment \$2,160
   Clothes \$3,600
   Medical \$4,320
   Savings \$3,600
   Odds and Ends \$2,160
4. \$600
5. \$690
6. \$63,000
7. Housing 32%
   Transportation 16%
   Food 14%
   Clothing 5%
   Health Care 6%
   Insurance and Pensions 9%
   Entertainment 6%
   Other 12%
8. 62%
9. \$2,520
10. \$1,680

**Pages 168–169**

1. **a.** \$171.72
   **b.** \$195.90
   **c.** \$24.18
2. **a.** \$95.23
   **b.** \$132
   **c.** \$36.77
3. **a.** \$418.70
   **b.** \$479.25
   **c.** \$60.55
4. **a.** \$481.95
   **b.** \$582.72
   **c.** \$100.77
5. **a.** \$307.35
   **b.** \$360.39
   **c.** \$53.04
6. **a.** \$240.45
   **b.** \$336.64
   **c.** \$96.19

**Pages 170–174, Posttest B**

1. **c.** $\frac{2}{3}$
2. **a.** $16\frac{17}{48}$
3. **a.** greater than 1
4. **d.** $52\frac{1}{16}$ pounds
5. **c.** $4\frac{8}{15}$
6. **b.** $5\frac{1}{2}$ inches
7. **d.** $\frac{3}{8}$
8. **b.** $2\frac{3}{4}$
9. **d.** 14
10. **a.** $5\frac{1}{4}$ pounds
11. **c.** $\frac{4}{5}$
12. **d.** $2\frac{2}{5}$
13. **d.** 200 bags
14. **a.** 320
15. **c.** 7.051
16. **b.** 0.004, 0.06, 0.064, 0.4
17. **c.** 26.392
18. **b.** 439 million
19. **d.** 12.214
20. **a.** 2.57
21. **b.** 0.086 meter
22. **a.** 54.384
23. **c.** 24.4
24. **d.** \$8.47
25. **c.** 4.3
26. **b.** 4
27. **c.** 6.54
28. **d.** 58 mph
29. **c.** 0.026
30. **a.** $42\frac{6}{7}\%$
31. **d.** $\frac{5}{16}$
32. **c.** 294
33. **d.** 210
34. **c.** $180 \div 6$
35. **a.** $87\frac{1}{2}\%$
36. **d.** 75%
37. **c.** 120
38. **b.** 280

# MEASUREMENTS AND FORMULAS

In this book you find *standard* or *customary* units of measurement such as feet and pounds. You also find *metric* units such as meters and kilograms. The list below shows common units of measurements and their equivalents in other units.

**Distance**

1 foot = 12 inches
1 yard = 3 feet
1 mile = 5,280 feet
1 mile ≈ 1.61 kilometers
1 inch = 2.54 centimeters
1 foot = 0.3048 meter
1 meter = 1,000 millimeters
1 meter = 100 centimeters
1 kilometer = 1,000 meters
1 kilometer ≈ 0.62 mile

**Area**

1 square foot = 144 square inches
1 square yard = 9 square feet
1 acre = 43,560 square feet

**Volume**

1 cup = 8 fluid ounces
1 quart = 4 cups
1 gallon = 4 quarts
1 gallon = 231 cubic inches
1 liter ≈ 0.264 gallon
1 cubic foot = 1,728 cubic inches
1 cubic yard = 27 cubic feet
1 board foot = 1 inch by 23 inches by 23 inches

**Weight/Mass**

1 ounce ≈ 28.350 grams
1 pound = 16 ounces
1 pound ≈ 453.592 grams
1 milligram = 0.001 grams
1 kilogram = 1,000 grams
1 kilogram ≈ 2.2 pounds
1 ton = 2,000 pounds

**Formulas**

| | |
|---|---|
| Perimeter ($P$) of a rectangle | $P = 2l + 2w$, where $l$ is the length and $w$ is the width |
| Area ($A$) of a rectangle | $A = lw$, where $l$ is the length and $w$ is the width |
| Volume ($V$) of a rectangular solid | $V = lwh$, where $l$ is the length, $w$ is the width, and $h$ is the height |
| Circumference ($C$) of a circle | $C = \pi d$, where $\pi \approx 3.14$ or $\frac{22}{7}$ and $d$ is the diameter |
| Area ($A$) of a circle | $A = \pi r^2$, where $\pi \approx 3.14$ or $\frac{22}{7}$ and $r$ is the radius |
| Interest ($i$) | $i = prt$, where $p$ is the principal, $r$ is the rate, and $t$ is the time |

# GLOSSARY

## A

**addition** The mathematical operation used to find a sum. The problem $0.2 + 1.3 = 1.5$ is an example.

**approximate** Another word for estimate.
As an adjective: close or almost exact. The symbol $\approx$ means "is approximately equal to."

**area** A measure of the amount of surface on a flat figure. A table top that is 6 feet long and 2.5 feet wide has an area of $6 \times 2.5 = 15$ square feet.

**average** A sum divided by the number of items that make up the sum. An average, also called the *mean*, is a representative number for a group. If one package weighs 2.3 ounces and another weighs 4.5 ounces, their average weight is the sum $(2.3 + 4.5 = 6.8)$ divided by 2 (the number of packages).
$6.8 \div 2 = 3.4$ ounces

## B

**borrowing** Regrouping the parts of the top number of a subtraction problem. In the problem $7 - 2\frac{3}{4}$, the 7 becomes $6\frac{4}{4}$.

## C

**canceling** Dividing a numerator in one fraction and a denominator in another fraction in a multiplication problem. For example, $\frac{4}{5} \times \frac{1}{6} = \frac{2}{15}$

**carrying** Regrouping the digits in an addition or multiplication problem. In the problem $2.6 + 1.9$, the sum of the tenths column is 15. The digit 5 remains in the tenths column, and the digit 1 is added to the units column.

**circle** A flat figure for which every point is the same distance from the center.

**circumference** A measure of the distance around a circle.

**common denominator** A number into which every denominator in a problem can divide evenly. For the problem $\frac{2}{3} + \frac{3}{4} + \frac{5}{6}$, the numbers 12, 24, and 36 are all common denominators.

**consecutive** One after the other. The numbers 9 and 10 are consecutive; so are 20 and 21.

## D

**decimal** A fraction in which the whole is divided into tenths, hundredths, thousandths, ten-thousandths, and so on. The decimal $0.7 = \frac{7}{10}$.

**decimal place** The position of a digit to the right of the decimal point. In the number 2.358, the digit 5 is in the hundredths place.

**decimal point** A dot that separates whole numbers from decimal fractions. In the number 14.6, the point separates the whole number 14 from the decimal fraction $\frac{6}{10}$.

**denominator** The bottom number or divisor in a fraction. In the fraction $\frac{3}{4}$, the denominator is 4.

**diameter** The distance across a circle.

**difference** The answer to a subtraction problem. For the problem $2.9 - 0.5$, the difference is 2.4.

**digit** One of the ten number symbols. The digits are 1, 2, 3, 4, 5, 6, 7, 8, 9, and 0.

**dividend** The number in a division problem into which another number divides. In $6.8 \div 2 = 3.4$, the dividend is 6.8.

**division** A mathematical operation that requires figuring out how many times one amount is contained in another. In the problem $\frac{3}{4} \div \frac{1}{8} = 6$, the answer means that there are exactly six one-eighths in $\frac{3}{4}$.

**divisor** The number in a division problem that divides into another. In $4.5 \div 0.5 = 9$, the divisor is 0.5.

## E

**estimate** As a noun: an approximate value. For the problem $2\frac{7}{8} + 3\frac{8}{9}$, a reasonable estimate of an answer is 7. As a verb: to find an approximate value.

## F

**fixed-place accuracy** In a division problem, rounding an answer to a certain place. The division problem $2 \div 0.3$ never comes out even. To the nearest hundredth, the answer is 6.67.

**formula** A mathematical rule written with an = sign. The formula for finding interest is $i = prt$, where $i$ is the interest, $p$ is the principal, $r$ is the rate, and $t$ is the time.

**fraction** A part of a whole. The fraction $\frac{1}{4}$ tells what part a quarter is of a dollar. A fraction can also be thought of as a division problem. The fraction $\frac{6}{2}$ means six divided by two.

**front-end rounding** Rounding the left-most digit of each number in a problem in order to calculate an estimate. In the problem $52 \times 0.79$, the number 52 rounds to 50 and 0.79 rounds to 0.8. The estimate is $50 \times 0.8 = 40$.

## H

**height** The straight-line measurement from the base of an object to the top. The measurements of a rectangular container include the length, the width, and the height.

**higher terms** The opposite of the reduced form of a fraction. The fraction $\frac{3}{4}$ raised to twelfths is $\frac{9}{12}$.

## I

**improper fraction** A fraction in which the numerator is equal to or larger than the denominator. The fractions $\frac{5}{5}$ and $\frac{4}{3}$ are both improper fractions.

**inequality** An expression that two amounts are not equal. The symbol < means *less than*. For example, $\frac{3}{8} < \frac{1}{2}$. The symbol > means *greater than*. For example, $\frac{5}{6} > \frac{1}{2}$.

**interest** A charge for borrowing money, usually calculated as a percent of the money borrowed. For example, an 8% interest charge on $1,000 is $80.

**invert** To reverse the position of the parts of a fraction. When the fraction $\frac{2}{3}$ is inverted, it becomes $\frac{3}{2}$.

## K

**kilogram** The standard unit of weight in the metric system. A kilogram is a little more than two pounds.

## L

**label** A word or abbreviation used to identify the unit of measurement of some quantity. An envelope has a weight in ounces. The label is ounces.

**liter** The standard unit of liquid measure in the metric system. A liter is about the same as one quart.

**lowest common denominator** The lowest number into which every denominator in a problem can divide evenly. For the problem $\frac{5}{6} + \frac{3}{4}$, the lowest common denominator is 12.

**lowest terms** The most reduced form of a fraction. The fraction $\frac{16}{20}$ reduced to lowest terms is $\frac{4}{5}$.

## M

**mean** Another word for average. A sum divided by the number of items that make up the sum.

**measurement** A dimension, quantity, or capacity. The measurements of a room usually include the length, the width, and the height.

**meter** The standard unit of length in the metric system. A meter is a little more than one yard.

**metric system** A standard of measure based on tens, hundreds, and thousands. The standard unit of length in the metric system is the meter. The standard unit of weight is the kilogram. The standard unit of liquid measure is the liter.

**mixed decimal** A number with both a whole number and a decimal fraction. In the mixed decimal 4.3, the whole number is 4 and the decimal fraction is 3 tenths.

**mixed number** A number with both a whole number and a fraction. For example, $5\frac{1}{2}$ is a mixed number.

**multiplication** A mathematical operation with whole numbers that consists of adding a number a certain number of times. For example, the problem $4 \times 3 = 12$ means finding the sum of three fours: $4 + 4 + 4 = 12$. With fractions, multiplication means finding a *part* of another number. For example, the problem $\frac{1}{3} \times 15 = 5$ means finding a third of fifteen.

## N

**numerator** The top number in a fraction. In the fraction $\frac{2}{9}$, the numerator is 2.

## O

**operation** A mathematical process such as addition, subtraction, multiplication, or division.

## P

**parallel** Being an equal distance apart. The sides opposite each other in a rectangle are parallel.

**percent** One hundredth or $\frac{1}{100}$. The symbol % means percent. An entire amount is 100%. Half of an entire amount is 50%.

**perimeter** A measure of the distance around a flat figure. A rectangle that is 5 meters long and 2.5 meters wide has a perimeter of 15 meters.

**place value** The number that every digit stands for. In the number 26.8, the digit 2 stands for 20 because 2 is in the tens place. The digit 8 stands for $\frac{8}{10}$ because 8 is in the tenths place.

**principal** The amount of money on which interest is calculated.

**product** The answer to a multiplication problem. For the problem $3.4 \times 2 = 6.8$, the product is 6.8.

**proper fraction** A fraction in which the numerator (the top number) is less than the denominator (the bottom number). The fractions $\frac{8}{9}$ and $\frac{1}{500}$ are both proper fractions.

## Q

**quotient** The answer to a division problem. In the problem $8.2 \div 2 = 4.1$, the quotient is 4.1.

## R

**radius** Half the diameter of a circle. Also, the distance from the center of the circle to any point on the circle.

**rate** An amount whose unit of measure contains a word such as *per* or *for each*. For example, the speed (or rate of speed) of a moving vehicle is often measured in *miles per hour*. Interest is measured in *percent*.

**reciprocal** The inverse of a fraction. $\frac{4}{5}$ is the reciprocal of $\frac{5}{4}$.

**rectangle** A four-sided figure with four square corners. A page of a textbook is a rectangle.

**rectangular solid** A three-dimensional figure such as a box that has length, width, and height.

**reducing** Expressing a fraction with smaller numbers. The fraction $\frac{9}{18}$ can be reduced to $\frac{1}{2}$. The reduced form of a fraction has the same value as the original form.

**rounding** Making an estimate that is close to an original amount. 4.862 rounded to the nearest tenth is 4.9.

## S

**square** A four-sided flat figure with four right angles and four equal sides.

**subtraction** The mathematical operation used to find the difference between two numbers. The problem $\frac{5}{8} - \frac{4}{8} = \frac{1}{8}$ is an example.

**sum** The answer to an addition problem. The sum of 0.3 and 0.4 is 0.7.

**symbol** A printed sign that represents an operation, a quantity, or a relationship. The symbol + means to add. The symbol ° means degree. The symbol < means less than.

## T

**time** A measurement from a point in the past to a more recent point. The units of measurement for time are seconds, minutes, hours, days, weeks, months, years, and so on.

**total** Another word for sum; the answer to an addition problem.

## U

**units of measurement** Labels for determining a quantity. For example, pounds and kilograms are units of measurement for weight.

**unit** The name of the right-most place in a whole number. In the number 36.2051, the digit 6 is in the units place.

## V

**volume** The amount of space occupied by a three-dimensional object. A rectangular box that is 4 feet long, 3 feet wide, and $\frac{1}{2}$ foot high has a volume of 6 cubic feet.

## W

**whole number** A number that can be divided evenly by 1. For example, 2 and 30 and 1,000 are all whole numbers.

## Z

**zero** The mathematical symbol 0. A fraction in which the numerator is much smaller than the denominator is close to 0 in value. For example, the fraction $\frac{1}{99}$ is much closer to 0 than the fraction $\frac{98}{99}$.

# INDEX